HORTICULTURE
Farm Development

HORTICULTURE
Farm Development

by

Bahaar Solanki | Laila Singh | Gauhar Chaudhary

Wordpen Academics
Publishing Science That Matter

ISBN: 978-93-6084-932-0

Published by : **WORDPEN ACADEMICS**
4736/23, Ansari Road, Darya Ganj,
New Delhi – 110 002
Phone: +91-011-35004336
E-mail: info@wordpenacademics.com

Printed at **:** **Replika Press Pvt. Ltd.**

Exclusive Distributor: ***Bio-Green Books, New Delhi***

PREFACE

"Horticulture Farm Development" embarks on a journey into the vibrant world of horticulture, offering readers a comprehensive guide to the principles, practices, and strategies for establishing and managing successful horticultural enterprises. This book serves as an invaluable resource for farmers, entrepreneurs, policymakers, and agricultural professionals seeking to harness the potential of horticulture to drive economic growth, enhance food security, and promote environmental sustainability.

Horticulture plays a vital role in global agriculture, encompassing the cultivation of fruits, vegetables, ornamental plants, medicinal herbs, and more. With its diverse range of crops and production systems, horticulture offers immense opportunities for diversification, value addition, and market expansion, thereby contributing to the resilience and viability of agricultural systems worldwide.

As editors of this volume, we have endeavored to compile a wealth of knowledge, expertise, and best practices from leading experts and practitioners in the field of horticulture. Through a blend of theoretical insights, practical methodologies, and case studies, this book covers a wide range of topics, including site selection and preparation, crop selection and management, irrigation and fertilization techniques, pest and disease management, post-harvest handling, marketing strategies, and farm business management.

Our overarching goal in presenting this book is to empower stakeholders across the horticulture value chain to unlock the full potential of horticulture as a driver of rural development, economic prosperity, and environmental stewardship. By providing practical guidance, innovative solutions, and real-world examples, we aim to inspire and support the establishment and growth of horticultural enterprises that contribute to food security, livelihoods, and sustainability in communities around the world.

We extend our heartfelt gratitude to all the contributors who have generously shared their expertise, experiences, and insights in the field of horticulture farm development. It is our hope that this volume will serve as a catalyst for innovation, collaboration, and transformative change in the horticulture sector, as we work together to build a more resilient, equitable, and sustainable future for all.

Bahaar Solanki | Laila Singh | Gauhar Chaudhary

CONTENTS

1

INTRODUCTION

Horticulture can be defined as the science, art, technology and business of cultivating ornamental plants and crops. Now, we have reached a point in time where we are more disconnected from nature than ever before, through urbanisation and the increased use of technology, perhaps it is now that society has the most to benefit from horticulture. This article will argue that horticulture has the potential to counter balance the negative impacts of modern living.

Horticultural crops are different; they incorporate yearly and perennial species, scrumptious natural products and vegetables, and enriching indoor and scene plants. These specialty crops offer assistance to maintain and enhance our lives by giving nutritious nourishment, improving the beauty of our homes and communities, and decreasing our carbon footprint.

Presently, we have come to a point in time where we are more detached from nature than ever before, through urbanization and the expanded utilization of innovation, perhaps it is now that society has the foremost advantage from horticulture.

Biophilia theory shows that exposure to nature is an inborn prerequisite for effective human conduct and advancement. This recommends that green scenes can offer assistance to supply a cure to the progressively urbanized and mechanical, modern way of living. The studies and researches have shown that the person who spends most of their time in timberlands is noted with a reduction in adrenaline hormone. Stress can cause mental fatigue and physiological harm to the body, therefore reducing it can play a crucial part in expanding wellbeing and wellbeing.

Production horticulture gives more coordinate and measurable benefits for health and prosperity, including extricates and oils for aromatherapy and homeopathy, nourishment, and others produce with various dietary and restorative benefits. Plants presently make up more than 25% of drugs in clinical use within the world, and discoveries are continuing.

Horticulture is the art of cultivating plants in gardens to produce food and medicinal ingredients, or for comfort and ornamental purposes. Horticulturists are agriculturists who grow flowers, fruits and nuts, vegetables and herbs, as well as ornamental trees and lawns. The study and practice of horticulture have been traced back thousands of years. Horticulture contributed to the transition from nomadic human communities to sedentary, or semi-sedentary, horticultural communities. Horticulture is divided into several categories which focus on the cultivation and processing of different types of plants and food items for specific purposes. In order to conserve the science of horticulture, multiple organizations worldwide educate, encourage, and promote the advancement of horticulture. Some notable horticulturists include Luca Ghini and Luther Burbank.

Horticulture involves plant propagation and cultivation to improve plant growth, yields, quality, nutritional value, and resistance to insects, diseases, and environmental stresses. It also includes plant conservation, landscape restoration, soil management, landscape and garden design, construction and maintenance, and arboriculture. The word horticulture is modeled after agriculture; it derives from the Latin words hortus and cultura, which mean "garden" and "cultivation", respectively. In contrast to agriculture, horticulture does not include intensive crop farming and large-scale crop production or animal husbandry. Additionally, horticulture focuses on the use of small plots with a wide variety of mixed crops while agriculture focuses on one large primary crop at a time.

Types of horticulture

There are several major areas of focus within the science of horticulture. They include:

- Olericulture: the production of vegetables.
- Pomology, also called fruticulture: the production of fruits and nuts.
- Viticulture: the production of grapes (largely intended for winemaking).
- Floriculture: the production of flowering and ornamental plants.
- Turf management: the production and maintenance of turf grass for sports, leisure and amenity use.

- Arboriculture: the cultivation and care of individual trees, shrubs, vines, and other perennial woody plants, primarily for landscape and amenity purposes.
- Landscape horticulture: the selection, production and care of plants used in landscape architecture.
- Postharvest physiology: the management of harvested horticultural crops to retard spoilage while stored or transported.

HISTORY

The origins of horticulture lie in the transition of human communities from a nomadic lifestyle as hunter-gatherers to sedentary, or semi-sedentary, horticultural communities. In the Pre-Columbian Amazon Rainforest, natives used biochar to enhance soil productivity by smoldering plant waste. European settlers called this soil Terra Preta de Indio. In forest areas, such horticulture was often carried out in swiddens, or "slash and burn" areas. In pre-contact North America, the semi-sedentary horticultural communities of the Eastern Woodlands, who grew maize, squash, and sunflower, contrasted markedly with the nomadic hunter-gatherer communities of the Plains people. Mesoamerican cultures focused in the cultivating of crops on a small scale, such as the "milpa" or maize field, around their dwellings or in specialized plots which were visited occasionally during migrations from one area to the next. In Central America, Maya horticulture involved augmentation of the forest with useful trees such as papaya, avocado, cacao, ceiba and sapodilla. In the cornfields, multiple crops such as beans, squash, pumpkins and chili peppers were grown, and in some cultures, these crops were tended mainly or exclusively by women

Advantages of Horticulture

Horticulture enriches diets: Horticulture — specifically, growing fruits and vegetables — provides critical nutrients for a balanced diet. Diets low in fruits and vegetables contribute significantly to some of the world's most widespread and debilitating nutrient-related disorders.

Horticulture increases incomes: Farmers growing high-value crops, such as fruits, vegetables, flowers or herbs, consistently earn more than those growing other commodities. Horticulture can be an engine for agricultural and economic diversification Improving livelihoods — by increasing farmer profits and diversifying nutrient-rich diets — are major goals of the Horticulture Innovation Lab's research efforts around the world. The program is guided by the Global Horticulture Assessment, an in-depth, collaborative, global analysis that also identified these challenges for horticulture development:

Gender equity: Vegetables, fruits and cut flowers are often grown and marketed by women, but women often have less access to markets, land, inputs and education. Addressing these constraints places women growers on the path to increasing productivity and expanding horticultural markets.

Technological innovation: Given the complexity of horticulture, innovative "leapfrog" technologies can reduce constraints and input costs that limit the ability of smallholder farmers to achieve maximum profitability.

Access to information and research capacity: Commercial success in horticulture depends on locally adapted research on tools such as improved cultivars, management tools, market knowledge and effective postharvest practices. Sustained horticultural growth requires access to reliable information, a well trained workforce and local capacity to conduct both original and adaptive research.

These three themes of gender equity, technological innovation, and information access are critical themes in all projects of the Horticulture Innovation Lab.

Improved Nutrition with the study of Horticulture, the crossbred species have led to an increase in the nutritional value of the fruits and vegetables.

Increases Resistance numerous crops are lost to bugs or severe climate. Another advantage of horticulture is that strains of plants that are safe from bugs or climate changes can be cultivated and developed. These strains could also be crossbred with other strains for improved flavor or nourishment or speedier growing times. Breeding plants with increased resistance to creepy crawlies and other pests implies fewer pesticides are utilized on the crops amid growth.

THE FINANCIAL BENEFITS OF HORTICULTURE

Green infrastructure provides healthy environments, which in-turn can contribute to improved socio-economic benefits. There is a good body of evidence to suggest that the quality of green space can impact on land and property prices, furthermore, there is some case-study specific evidence that it can have a positive impact on economic factors, including new business start-ups and private investment. The markets available within the production sector are dynamic and diverse, including foods, ornamental plants, essential oils and pharmaceuticals, presenting a myriad of profit-making opportunities.

Green infrastructure in urban environments can reduce cooling energy requirements. Studies in North America have indicated that, even if a conservative reduction of 20% in energy use is applied, this would result in a US$10 billion saving per annum.

HORTICULTURE AND CULTURAL IDENTITY

Cultures across the world, throughout man's evolution, have co-existed with plants, providing food, raw materials, shelter, shade and religious symbolism and have become integral to local identity and character. The importance of this relationship is highlighted in Italy where Olive trees (Olea europapaea), central to the identity and heritage of the region, are currently under threat from Xylella fastidiosa. Plants are used on national flags (for example, the Canadian flag with a maple leaf), or as national flowers, and for special occasions such as weddings and funerals.

HABITAT, HERITAGE AND BIODIVERSITY

Culture, myths and people, as much as plants and hard landscaping, create the genius loci or 'spirit of place' associated with historic and heritage gardens, and this plays a central role in their preservation and maintenance. Wild landscapes are decreasing with growing human population, therefore human-managed land will play an increasingly important role in biodiversity preservation on a global scale. A study by Baillie and Gregory in 1998 indicated that human-made environments hold greater than 20% of British populations of blackbirds, song thrushes and starlings. Gardens provide a vast array of important habitats for wildlife including: ponds that support amphibians, garden plants that provide nectar for butterflies and bees, and woodpiles as hibernation sites (Royal Society for the Protection of Birds, n.d.).

Plant conservation takes place within gardens, for example, Brugmansia sanguinea found in gardens, is now lost to the wild. Seed banks such as the Kew Gardens' Millennium Seed Bank, which can potentially facilitate future ecological restoration projects, are therefore providing a genetic insurance policy for future generations. Organisations such as Botanic Gardens Conservation International and Plant Heritage play crucial roles in conservation on a national and international level. Also, the financial drive to produce new horticultural crops and ornamental plants is ultimately increasing biological diversity in cultivated plant communities.

THE VISUAL BENEFITS OF HORTICULTURAL LANDSCAPES

Plants, individually or collectively, are considered to be attractive, possessing some combination of symmetry, shape, texture, colour, form, fragrance and taste, furthermore, green is psychologically the most restful colour. Wider notions of what is thought to be aesthetically pleasing in horticulture, such as style, are subjective and open to individual interpretation based on cultural and individual inclinations. The late Sir Geoffrey Jellicoe argues that landscape aesthetics must appeal to the evolutionary and collective subconscious, for instance, the appeal of parkland and open spaces derives from our time as hunter gatherers, and the attraction to the formality in straight lines, such as a neatly clipped hedge, stems from the birth of agriculture. Contemporary horticulture incorporates naturalism, which is considered to have originated in China, and although naturalism is normally less tidy, it is thought that knowledge of the ecological function of a "messy" garden can increase its aesthetic value.

'Wabi Sabi', a Japanese word describing that which is imperfect, transient, impermanent and a symbiosis of the work done by nature and humans, perhaps offers a deeper insight in to the aesthetic that humans find attractive in gardens. It is described by the narrative: a Japanese master had his son clean, tidy and scrub the tea ceremony yard for three days until it was perfectly clean. The master then shook an autumnal maple tree branch, so the leaves fell randomly to the floor, letting nature give the final touch of artistry.

DISADVANTAGES OF HORTICULTURE

Smaller Yields

Horticulture tends to include smaller crops than agribusiness since it is harder to control and keep up growing conditions on a bigger scale. As a result, the crops yield less product than conventional agribusiness crops. The perfect environment for a horticulture crop is an encased plant where the horticulturist can control all viewpoints of the developing, counting soil substance, water recurrence, and quality and bug control.

Reduced Soil Quality

While crops developed utilizing horticultural strategies are normally very nutritious and developed in soil containing the proper composition of supplements, the soils used are rapidly drained of their nutrients. The ash from the natural material adds supplements and minerals to the soil, acting as manure.

It requires high maintenance and a huge labor force.

Horticulture Crops produced in India

The diversity of physiographic, climatic, and soil characteristics empowers India to develop a huge variety of horticultural crops. India come up as the world's biggest producers of vegetables.

Orange

Or Mandarin orange grows effectively in tropical and subtropical parts basically under rain-fed conditions, in Coorg, Wynaad tract, Palani slopes, and the Nilgiris within the south between elevations of 600 and 1,500 m. Orange can be developed effectively on a wide range of soils, but the perfect soil is medium or light soil with a slightly heavier subsoil. Heavy dark soil, underlain with a murrain and having great drainage, is additionally suitable. In the slopes and sticky locales where planting is generally done on steep slants, the land is properly terraced. Within the fields, where the trees have to be watered, the land ought to be leveled. The places of production of oranges are Assam, Nagpur, Punjab, Wynand, Coorg, Palani slopes, and the Nilgiris.

Mango

About 50 percent of the total zone under natural products in the nation is possessed by mango. The mango tree develops all through the nation from the ocean level up to approximately 1,500 m. It is versatile to a wide range of soil and climatic conditions. It can withstand both dry conditions and heavy precipitation. The mangos is a native of rainstorm lands and favors a climate with 75-250 cm precipitation concentrated from June to September and mean shade temperatures of approximately 28 °C. The greatest generation of mangoes in India comes from Uttar Pradesh, Bihar, Andhra Pradesh, Maharashtra, Tamil Nadu, and Kerala among others.

Grapes

Being a subtropical fruit grows well in dry climate having a brief sharp winter and a long dry summer. They don't flourish in districts having muggy summers. Grape develops best on light, friable, loamy soils with free seepage. Heavy soils are unsuitable. Grapes are mainly grown in Punjab, Uttarakhand, Himachal Pradesh, Maharashtra.

Guava

Could be an exceptionally hard tree that can withstand warm and droughts? A cool winter induces heavy fruiting. The whole area beneath guava is almost 30,000 hectare and Uttar Pradesh has the biggest area under guava taken after by Bihar.

Pineapple

It is a humid tropical plant and grows well both within the fields and also at elevations not surpassing 900 m. It develops in all sorts of soils but can endure neither very high temperature nor frost. Assam, Meghalaya, West Bengal, Tripura, Uttar Pradesh, Andhra Pradesh, Kerala, and Karnataka are the imperative places of production of pineapple.

ECOSYSTEM SERVICES AND HORTICULTURE IN URBAN AREAS

Benefits that are provided to humans by natural ecosystems are called ecosystem services; They include factors such as regulating climate, soil formation, oxygen production and recreational benefits. As the growing global population is placing an increasing strain on the world's natural systems, horticultural landscapes can can help to provide some of these benefits that have been lost through degradation and loss of wild environments. In particular carbon sequestration, which can help to regulate climate, can be optimised with planting that requires minimum soil disturbance. This way the soil is more able to become a sink for carbon storage as the decomposition of organic matter is slowed.

Within urban environments planting and green infrastructure can moderate temperatures in both winter and summer through shading, evapotranspiration and the reduction of air movement. Furthermore, there is strong evidence to suggest that through interception, infiltration and encouraging better water absorption into the soil, green infrastructure attenuates storm and flood waters, and its associated pollution. This, along with sustainable urban drainage systems can help with ground water recharge. There is also some partial evidence to suggest that certain plants can screen out aerial particulate matter and improve air quality.

HORTICULTURE - HEALTH AND WELLBEING

Natural ecosystems, as well as horticultural landscapes, can help to sustain and improve human health and wellbeing. Biophilia hypothesis indicates that exposure to nature is an intrinsic requirement for effective human behaviour and development. This suggests that green landscapes can help to provide an antidote to the increasingly urbanised and technological, contemporary way of living. A recent study about people who had spent time in forest environments in Japan, and illustrated in Figure 1, showed this restoration from stress translates, in part, as a reduction in adrenaline. Stress can cause mental fatigue and physiological damage to the body, therefore reducing it can play a vital role in increasing health and wellbeing.

The reduced risk of diseases associated with time spent in green space has been attributed to the increased likelihood of participating in sustained exercise, and bringing people together to create social cohesion and strong social bonds. Other benefits of green infrastructure include: providing a physical barrier to wind and noise and mitigating urban temperatures through shading and evapotranspiration.

The list of recorded psychophysiological benefits that green space provides also includes: improved cognitive function, self-discipline and alleviation of attention deficit disorder symptoms in children. Indeed, nature-based learning can increase child development through, amongst other things, exploration and discovery, especially boosting self-esteem and sense of self.

Production horticulture provides more direct and measurable benefits for health and wellbeing, including: extracts and oils for aromatherapy and homeopathy, food, and other produce with numerous nutritional and medicinal benefits. Plants now make up more than 25% of drugs in clinical use in the world, and new discoveries are continuing: recently a treatment for ovarian cancer was discovered in the bark of pacific yew (Taxus brevifolia).

Physical and Nutritional Benefits of Consumer Horticulture

The busy hustle and bustle of everyday life makes it a challenge to balance professional demands and family obligations with physical activity and planned and healthy diets. Even though numerous studies have linked regular physical activity with improvements in the function of muscles and joints, good bone mass, improved metabolism, immunity to disease, and enhanced mental health, it is still a challenge to ?t in regular physical activity on a daily basis. It would be great to incorporate these things into one activity. Gardening can do just that, meeting our needs for physical, nutritional, mental, and social health and well-being.

Gardening offers exercise for upper and lower body strength:

- Digging
- Weeding
- Planting
- Raking
- Staking plants

Gardening activities are well suited to meet recommendations for daily physical activity, such as those recommended by the American Heart Association (AHA). Gardening tasks using both upper and lower body strength, such as digging, fertilizing, weeding, raking, and tying plants to stakes, offer moderate-intensity physical activity, comparable to a brisk walk, swimming, dancing, and biking. Other gardening tasks that use the upper body strength while standing or squatting, such as pruning, mixing soil, planting seedlings, sowing, watering, and harvesting, are low-intensity physical activities, as are filling containers with soil and washing harvested produce. Together, these activities satisfy AHA's weekly physical activity guidelines, such as 150 minutes of moderate-intensity aerobic activity, 75 minutes of vigorous aerobic activity, or a combination of both for adults. For children ages 6 to 17, gardening can meet the recommended 60 minutes per day of moderate-intensity aerobic activity and strengthening activities at least three days per week.

Gardening can also contribute to good health by providing healthy fruits and vegetables for adequate nutrition. When a household grows fresh fruits and vegetables, the entire household tends to eat more fruits and vegetables. Children who participate in growing vegetables are more likely to try or eat a broader selection of vegetables.

It is well-known that plants are an excellent source of fiber, vitamins, and minerals essential to health. Many plants have been used not only for food, but also for healing. Most drugs used today have roots in plants, pun intended. Herbal medicines have been used by humans throughout our natural history. Pharmaceutical companies utilize plant compounds to develop new therapeutic drugs and have derived many medicines, such as aspirin, anti-inflammatory, anti-diabetic compounds, and chemotherapeutic compounds.

Activities like gardening help us remain active and help make healthy food choices easier so that we minimize our risk for diseases, such as diabetes, heart disease and stroke, osteoarthritis, and many types of cancers that have been directly linked to obesity that has reached epidemic proportions globally. Growing a vegetable garden or tending gardens and landscapes offers physical activity and nutritious, low-calorie foods essential for maintaining a healthy weight. Benefits are not limited to food gardening, but also extend to any type of gardening that involves physical activity.

Gardening keeps us active and healthy so that we minimize our risk for chronic diseases.

Mental and Emotional Benefits of Consumer Horticulture

In addition to physical and nutritional health benefits, gardening uniquely provides important mental well-being benefits. The repetitive tasks associated with gardening are reported to be relaxing and o?er mental restoration and focus recovery. Many gardeners report "stress relief" and therapeutic benefits resulting from the emotional processing that comes while tending gardens and plants. Gardeners listed mental benefits from gardening such as a sense of purpose, relaxing, and forgetting worries. Studies have shown that exposure to plants while recovering from surgery or being in the hospital has improved recovery. Gardening can enhance the ability to respond and rebound after dificulties, such as stress or illness.

Gardens give us pride and a sense of purpose.

Gardening can meet our need for personal choice, mastery and competence, and a sense of relationship to others. Designing a garden space, selecting which plants to grow or even their flower color are examples of personal choice. Selecting a preferred garden tool may meet that need for a child in the garden. Growing edible plants or those that are purely ornamental is entirely up to the gardener.

Opportunities to achieve success and demonstrate competency abound in the garden. A sense of pride and purpose results from growing and harvesting that ?rst radish or tomato or from arranging a vase of cut flowers. Though success in the garden leads to an increase in self-esteem for both adults and youth, they should not worry about straight lines or perfection in the garden, or even plant death. Plants teach youth and adults alike about patience, delayed gratification, and nurturing. The garden and its related tasks provide opportunity to teach work habits by helping in the garden and understanding what it takes to produce food.

Garden tasks, such as watering and fertilizing our plants or those of a neighbor, help us observe growth and realize what it means to be needed. Tending the garden offers valuable focused time for our families to converse, share, and interact, supporting feelings of relatedness and belongingness and creating family traditions. A family's gardening example often encourages other families to start gardens, and sharing knowledge and experiences encourages fellowship.

Gardens provide safe areas for exchange and social events, such as conversing with neighbors or even having a neighborhood picnic. Creating personal connections through gardening can help reinforce healthy food choices and regular physical activity. Community benefits, such as a sense of mutual appreciation and support, pride of place, and development of social networks, often result from gardening, offering opportunities for neighbors to connect. The garden offers a place for newcomers to settle in and establish a new home or place of their own that reflects their cultural practices, such as specific plants and foods that remind them of home.

Growing plants can make our neighborhoods safer places to live. Routinely taking care of plants gets us outside, talking with others, and becoming more aware of our neighborhood, resulting in an increased sense of community, stronger social connections, and reduced crime, such as less graffti, litter, and verbal aggression among residents. The plants we cultivate in the places we live send a message that someone cares and that a property is valued. The plants and gardens that we tend matter for our personal health. They can enhance our homes, making them a haven for rest and well-being. They provide much-needed exercise, nutrition, and diversion while also providing the space to make important community connections.

Horticultural regions

Temperate zones

Temperate zones for horticulture cannot be defined exactly by lines of latitude or longitude but are usually regarded as including those areas where frost in winter occurs, even though rarely. Thus, most parts of Europe, North America, and northern Asia are included, though some parts of the United States, such as southern Florida, are considered subtropical. A few parts of the north coast of the Mediterranean and the

Mediterranean islands are also subtropical. In the Southern Hemisphere, practically all of New Zealand, a few parts of Australia, and the southern part of South America have temperate climates. For horticultural purposes altitude is also a factor; the lower slopes of great mountain ranges, such as the Himalayas and the Andes, are included. Thus, the temperate zones are very wide and the range of plants that can be grown in them is enormous, probably greater than in either the subtropical or tropical zones. In the temperate zones are the great coniferous and deciduous forests: pine, spruce, fir, most of the cypresses, the deciduous oaks (but excluding many of the evergreen ones), ash, birch, and linden.

The temperate zones are also the areas of the grasses—the finest lawns particularly are in the regions of moderate or high rainfall—and of the great cereal crops. Rice is excluded as being tropical, but wheat, barley, corn (maize), and rye grow well in the temperate zones.

Plants in the temperate zones benefit from a winter resting season, which clearly differentiates them from tropical plants, which tend to grow continuously. Bulbs, annuals, herbaceous perennials, and deciduous trees become more frost-resistant with the fall of sap and therefore have a better chance of passing the resting season undamaged. Another influence is the varying length of darkness and light throughout the year, so that many plants, such as chrysanthemums, have a strong photoperiodism. The chrysanthemum flowers only in short daylight periods, although artificial lighting in nurseries can produce flowers the year round.

Most of the great gardens of the world have been developed in temperate zones. Particular features such as rose gardens, herbaceous borders, annual borders, woodland gardens, and rock gardens are also those of temperate-zone gardens. Nearly all depend for their success on the winter resting period.

Tropical zones

There is no sharp line of demarcation between the tropics and the subtropics. Just as many tropical plants can be cultivated in the subtropics, so also many subtropical and even temperate plants can be grown satisfactorily in the tropics. Elevation is a determining factor. For example, the scarlet runner bean, a common plant in temperate regions, grows, flowers, and develops pods normally on the high slopes of Mount Meru in Africa near the Equator, but it will not set pods in Hong Kong, a subtropical situation a little south of the Tropic of Cancer but at a low elevation.

In addition to elevation, another determinant is the annual distribution of rainfall. Plants that grow and flower in the monsoon areas, as in India, will not succeed where the climate is uniformly wet, as in Bougainville in the Solomon Islands. Another factor is the length of day, the number of hours the Sun is above the horizon; some plants flower only if the day is long, but others make their growth during the long days and flower when the day is short. Certain strains of the cosmos plant are so sensitive to light that where the day is always about 12 hours, as near the Equator, they flower when only a few inches high; if grown near the Tropic of Cancer or the Tropic of Capricorn, they attain a height of several feet, if the seeds are sown in the spring, before flowering in the short days of autumn and winter. Poinsettia is a short-day plant that may be seen in flower in Singapore on any day of the year, while in Trinidad it is a blaze of glory only in late December.

In the tropics of Asia and parts of Central and South America, the dominant features of the gardens are flowering trees, shrubs, and climbers. Herbaceous plants are relatively few, but many kinds of orchids can be grown.

Vegetable crops vary in kind and quality with the presence or absence of periodic dry seasons. In the uniformly wet tropics, the choice is limited to a few root crops and still fewer greens. Sweet potatoes grow and bear good crops where the average monthly rainfall, throughout the year, exceeds 25 cm (10 inches), but they grow even better where there is a dry season. The same can be said of taro, yams, and cassava. Tropical greens from the Malay Peninsula are not as good as those grown in South China, the Hawaiian Islands, and Puerto Rico. They include several spinaches, of which Chinese spinach or amaranth is the best; several cabbages; Chinese onions and chives; and several gourds, cucumbers, and, where there is a dry season, watermelons. Eggplants, peppers, and okra are widely cultivated. Many kinds of beans can be grown successfully, including the French bean from the American subtropics, the many varieties of the African cowpea, and yard-long bean. The yam bean, a native of tropical America, is grown for its edible tuber. In the drier areas the pigeon pea, the soybean, the peanut (groundnut), and the Tientsin green bean are important crops. Miscellaneous crops include watercress, ginger, lotus, and bamboo.

Gardening

Gardening is the practice of growing and cultivating plants as part of horticulture. In gardens, ornamental plants are often grown for their flowers, foliage, or overall appearance; useful plants, such as root vegetables, leaf vegetables, fruits, and herbs, are grown for consumption, for use as dyes, or for medicinal or cosmetic use.

Gardening ranges in scale from fruit orchards, to long boulevard plantings with one or more different types of shrubs, trees, and herbaceous plants, to residential back gardens including lawns and foundation plantings, all the way to container gardens grown inside or outside. Gardening may be very specialized, with only one type of plant grown, or involve a variety of plants in mixed plantings. It involves an active participation in the growing of plants, and tends to be labor-intensive, which differentiates it from farming or forestry.

Ancient times

Forest gardening, a forest-based food production system, is the world's oldest form of gardening. Forest gardens originated in prehistoric times along jungle-clad river banks and in the wet foothills of monsoon regions. In the gradual process of families improving their immediate environment, useful tree and vine species were identified, protected and improved while undesirable species were eliminated. Eventually foreign species were also selected and incorporated into the gardens.

After the emergence of the first civilizations, wealthy individuals began to create gardens for aesthetic purposes. Ancient Egyptian tomb paintings from the New Kingdom (around 1500 BC) provide some of the earliest physical evidence of ornamental horticulture and landscape design; they depict lotus ponds surrounded by symmetrical rows of acacias and palms. A notable example of ancient ornamental gardens were the Hanging Gardens of Babylon—one of the Seven Wonders of the Ancient World —while ancient Rome had dozens of gardens.

Wealthy ancient Egyptians used gardens for providing shade. Egyptians associated trees and gardens with gods, believing that their deities were pleased by gardens. Gardens in ancient Egypt were often surrounded by walls with trees planted in rows. Among the most popular species planted were date palms, sycamores, fig trees, nut trees, and willows. These gardens were a sign of higher socioeconomic status. In addition, wealthy ancient Egyptians grew vineyards, as wine was a sign of the higher social classes. Roses, poppies, daisies and irises could all also be found in the gardens of the Egyptians.

Assyria was also renowned for its beautiful gardens. These tended to be wide and large, some of them used for hunting game—rather like a game reserve today—and others as leisure gardens. Cypresses and palms were some of the most frequently planted types of trees. Gardens were also available in Kush. In Musawwarat es-Sufra, the Great Enclosure dated to the 3rd century BC included splendid gardens.

Ancient Roman gardens were laid out with hedges and vines and contained a wide variety of flowers—acanthus, cornflowers, crocus, cyclamen, hyacinth, iris, ivy, lavender, lilies, myrtle, narcissus, poppy, rosemary and violets—as well as statues and sculptures. Flower beds were popular in the courtyards of rich Romans.

THE MIDDLE AGES

The Middle Ages represent a period of decline in gardens for aesthetic purposes. After the fall of Rome, gardening was done for the purpose of growing medicinal herbs and/or decorating church altars. Monasteries carried on a tradition of garden design and intense horticultural techniques during the medieval period in Europe. Generally, monastic garden types consisted of kitchen gardens, infirmary gardens, cemetery orchards, cloister garths and vineyards. Individual monasteries might also have had a "green court", a plot of grass and trees where horses could graze, as well as a cellarer's garden or private gardens for obedientiaries, monks who held specific posts within the monastery.

Islamic gardens were built after the model of Persian gardens and they were usually enclosed by walls and divided in four by watercourses. Commonly, the centre of the garden would have a reflecting pool or pavilion. Specific to the Islamic gardens are the mosaics and glazed tiles used to decorate the rills and fountains that were built in these gardens.

By the late 13th century, rich Europeans began to grow gardens for leisure and for medicinal herbs and vegetables. They surrounded the gardens by walls to protect them from animals and to provide seclusion. During the next two centuries, Europeans started planting lawns and raising flowerbeds and trellises of roses. Fruit trees were common in these gardens and also in some, there were turf seats. At the same time, the gardens in the monasteries were a place to grow flowers and medicinal herbs but they were also a space where the monks could enjoy nature and relax.

The gardens in the 16th and 17th century were symmetric, proportioned and balanced with a more classical appearance. Most of these gardens were built around a central axis and they were divided into different parts by hedges. Commonly, gardens had flowerbeds laid out in squares and separated by gravel paths.

Gardens in Renaissance were adorned with sculptures, topiary and fountains. In the 17th century, knot gardens became popular along with the hedge mazes. By this time, Europeans started planting new flowers such as tulips, marigolds and sunflowers.

Cottage gardens Cottage gardens, which emerged in Elizabethan times, appear to have originated as a local source for herbs and fruits. One theory is that they arose out of the Black Death of the 1340s, when the death of so many laborers made land available for small cottages with personal gardens. According to the late 19th-century legend of origin, these gardens were originally created by the workers that lived in the cottages of the villages, to provide them with food and herbs, with flowers planted among them for decoration. Farm workers were provided with cottages that had architectural quality set in a small garden—about 1 acre (0.40 ha)—where they could grow food and keep pigs and chickens.

Authentic gardens of the yeoman cottager would have included a beehive and livestock, and frequently a pig and sty, along with a well. The peasant cottager of medieval times was more interested in meat than flowers, with herbs grown for medicinal use rather than for their beauty. By Elizabethan times there was more prosperity, and thus more room to grow flowers. Even the early cottage garden flowers typically had their practical use—violets were spread on the floor (for their pleasant scent and keeping out vermin); calendulas and primroses were both attractive and used in cooking. Others, such as sweet William and hollyhocks, were grown entirely for their beauty.

18th century

In the 18th century gardens were laid out more naturally, without any walls. This style of smooth undulating grass, which would run straight to the house, clumps, belts and scattering of trees and his serpentine lakes formed by invisibly damming small rivers, were a new style within the English landscape, a "gardenless" form of landscape gardening, which swept away almost all the remnants of previous formally patterned styles. The English landscape garden usually included a lake, lawns set against groves of trees, and often contained shrubberies, grottoes, pavilions, bridges and follies such as mock temples, Gothic ruins, bridges, and other picturesque architecture, designed to recreate an idyllic pastoral landscape. This new style emerged in England in the early 18th century, and spread across Europe, replacing the more formal, symmetrical garden à la française of the 17th century as the principal gardening style of Europe. The English garden presented an idealized view of nature. They were often inspired by paintings of landscapes by Claude Lorraine and Nicolas Poussin, and some were Influenced by the classic Chinese gardens of the East, which had recently been described by European travelers. The work of Lancelot 'Capability' Brown was particularly influential. Also, in 1804 the Horticultural Society was formed.

Gardens of the 19th century contained plants such as the monkey puzzle or Chile pine. This is also the time when the so-called "gardenesque" style of gardens evolved. These gardens displayed a wide variety of flowers in a rather small space. Rock gardens increased in popularity in the 19th century. India: In India, in the ancient times, patterns from sacred geometry and mandalas were used to design their gardens. Distinct mandala patterns denoted specific deities, planets, or even constellations. Such a garden was also referred to as a 'Mandala Vaatika'. The word 'Vaatika' can mean garden, plantation or parterre.

Types

Residential gardening takes place near the home, in a space referred to as the garden. Although a garden typically is located on the land near a residence, it may also be located on a roof, in an atrium, on a balcony, in a window box, on a patio or vivarium.

Gardening also takes place in non-residential green areas, such as parks, public or semi-public gardens (botanical gardens or zoological gardens), amusement parks, along transportation corridors, and around tourist attractions and garden hotels. In these situations, a staff of gardeners or groundskeepers maintains the gardens.

Indoor gardening is concerned with the growing of houseplants within a residence or building, in a conservatory, or in a greenhouse. Indoor gardens are sometimes incorporated as part of air conditioning or heating systems. Indoor gardening extends the growing season in the fall and spring and can be used for winter gardening.

Native plant gardening is concerned with the use of native plants with or without the intent of creating wildlife habitat. The goal is to create a garden in harmony with, and adapted to a given area. This type of gardening typically reduces water usage, maintenance, and fertilization costs, while increasing native faunal interest.

Water gardening is concerned with growing plants adapted to pools and ponds. Bog gardens are also considered a type of water garden. These all require special conditions and considerations. A simple water garden may consist solely of a tub containing the water and plant(s). In aquascaping, a garden is created within an aquarium tank.

Container gardening is concerned with growing plants in any type of container either indoors or outdoors. Common containers are pots, hanging baskets, and planters. Container gardening is usually used in atriums and on balconies, patios, and roof tops.

Hügelkultur is concerned with growing plants on piles of rotting wood, as a form of raised bed gardening and composting in situ. An English loanword from German, it means "mound garden." Toby Hemenway, noted permaculture author and teacher, considers wood buried in trenches to also be a form of hugelkultur referred to as a dead wood swale. Hugelkultur is practiced by Sepp Holzer as a method of forest gardening and agroforestry, and by Geoff Lawton as a method of dryland farming and desert greening. When used as a method of disposing of large volumes of waste wood and woody debris, hugelkultur accomplishes carbon sequestration. It is also a form of xeriscaping.

Community gardening is a social activity in which an area of land is gardened by a group of people, providing access to fresh produce, herbs, flowers and plants as well as access to satisfying labor, neighborhood improvement, sense of community and connection to the environment. Community gardens are typically owned in trust by local governments or nonprofits.

Garden sharing partners landowners with gardeners in need of land. These shared gardens, typically front or back yards, are usually used to produce food that is divided between the two parties.

Organic gardening uses natural, sustainable methods, fertilizers and pesticides to grow non-genetically modified crops.

Biodynamic gardening or biodynamic agriculture is similar to organic gardening, but it includes various esoteric concepts drawn from the ideas of Rudolf Steiner, such as astrological sowing and planting calendar and particular field and compost preparations.

Commercial gardening is a more intensive type of gardening that involves the production of vegetables, nontropical fruits, and flowers from local farmers. Commercial gardening began because farmers would sell locally to stop food from spoiling faster because of the transportation of goods from a far distance. Mediterranean agriculture is also a common practice that commercial gardeners use. Mediterranean agriculture is the practice of cultivating animals such as sheep to help weed and provide manure for vine crops, grains, or citrus. Gardeners can easily train these animals to not eat the actual plant.

Social aspects

People can express their political or social views in gardens, intentionally or not. The lawn vs. garden issue is played out in urban planning as the debate over the "land ethic" that is to determine urban land use and whether hyper hygienist bylaws (e.g. weed control) should apply, or whether land should generally be allowed to exist in its natural wild state. In a famous Canadian Charter of Rights case, "Sandra Bell vs. City of Toronto", 1997, the right to cultivate all native species, even most varieties deemed noxious or allergenic, was upheld as part of the right of free expression.

Community gardening comprises a wide variety of approaches to sharing land and gardens.People often surround their house and garden with a hedge. Common hedge plants are privet, hawthorn, beech, yew, leyland cypress, hemlock, arborvitae, barberry, box, holly, oleander, forsythia and lavender. The idea of open gardens without hedges may be distasteful to those who enjoy privacy. The Slow Food movement has sought in some countries to add an edible school yard and garden classrooms to schools, e.g. in Fergus, Ontario, where these were added to a public school to augment the kitchen classroom. Garden sharing, where urban landowners allow gardeners to grow on their property in exchange for a share of the harvest, is associated with the desire to control the quality of one's food, and reconnect with soil and community.

In US and British usage, the production of ornamental plantings around buildings is called landscaping, landscape maintenance or grounds keeping, while international usage uses the term gardening for these same activities.

Also gaining popularity is the concept of "Green Gardening" which involves growing plants using organic fertilizers and pesticides so that the gardening process – or the flowers and fruits produced thereby – doesn't adversely affect the environment or people's health in any manner.

Benefits

Gardening is considered by many people to be a relaxing activity. There are also many studies about the positive effects on mental and physical health in relation to gardening. Specifically, gardening is thought to increase self-esteem and reduce stress. As writer and former teacher Sarah Biddle notes, one's garden may become a "tiny oasis to relax and recharge [one's] batteries."

Comparison with farming

Berms of fava beans have been planted at Hayes Valley Farm, a community-built farm on the former Central freeway ramps of San Francisco

Gardening for beauty is likely nearly as old as farming for food, however for most of history for the majority of people there was no real distinction since the need for food and other useful products trumped other concerns. Small-scale, subsistence agriculture (called hoe-farming) is largely indistinguishable from gardening. A patch of potatoes grown by a Peruvian peasant or an Irish smallholder for personal use could be described as either a garden or a farm. Gardening for average people evolved as a separate discipline, more concerned with aesthetics, recreation and leisure, under the influence of the pleasure gardens of the wealthy.[citation needed] Meanwhile, farming has evolved (in developed countries) in the direction of commercialization, economics of scale, and monocropping.

In respect to its food-producing purpose, gardening is distinguished from farming chiefly by scale and intent. Farming occurs on a larger scale, and with the production of salable goods as a major motivation. Gardening happens on a smaller scale, primarily for pleasure and to produce goods for the gardener's own family or community. There is some overlap between the terms, particularly in that some moderate-sized vegetable growing concerns, often called market gardening, can fit in either category.

The key distinction between gardening and farming is essentially one of scale; gardening can be a hobby or an income supplement, but farming is generally understood as a full-time or commercial activity, usually involving more land and quite different practices. One distinction is that gardening is labor-intensive and employs very little infrastructural capital, sometimes no more than a few tools, e.g. a spade, hoe, basket and watering can. By contrast, larger-scale farming often involves irrigation systems, chemical fertilizers and harvesters or at least ladders, e.g. to reach up into fruit trees.

However, this distinction is becoming blurred with the increasing use of power tools in even small gardens.

The term precision agriculture is sometimes used[by whom?] to describe gardening using intermediate technology (more than tools, less than harvesters), especially of organic varieties. Gardening is effectively scaled up to feed entire villages of over 100 people from specialized plots. A variant is the community garden which offers plots to urban dwellers; see further in allotment (gardening).

Indigenous horticulture

Indigenous horticulture is practised in various ways across all inhabited continents. Indigenous refers to the native peoples of a given area and horticulture is the practice of small-scale intercropping.

North Africa

In North Africa, one such example is the farming practices of the Eggon, a Nigerian hill farming community. The Eggon live in the Mada hills, between Lafia and Akwanga. The hills lay between two rivers, the Mada and Arikya. The altitude helps crops retain moisture on the hills, due to early morning mists and fogs; this also makes for earlier and longer crop cultivation. They practice bush fallow agriculture as well as mixed farming land management styles. They focus on growing yams, cassava, maize, beans, and African rice; much of what is produced is exported as a cash crop and is their primary source of cash income. The Eggon use a terraced agricultural system to maximize space on the hills. The goats they raise are kept mostly for fertilizers used in farming. They are only killed on special occasions, such as weddings. The Eggon use the diversity in their environment to maximize their crop production.

West Africa

In West Africa, the Kissidougou live on the savannah, dotted by dense areas of "forest islands" created by them. The Kissidougou practice intercropping within the forested areas. However, they also operate farms maintained on slopes or plateaus located between the forest islands. They prepare the savannah lands for forestation through farming and burning of the grasses to fertilize the soils. The Kissidougou graze their cattle on the savannah to help to maintain flammable grasses around the farms and the villages. The Kissidougou create diversity in their environment by farming and transforming savannah into lush, dense forest.

The prevalence of wetlands in West Africa has helped to support local indigenous horticulture. Seasonal flooding of major rivers in the region, such as the Niger, the Sudd, and the Senegal, have made it possible for flood-cropping in many areas. Indigenous people have utilized a variety of irrigation techniques in order to take advantage of this flooding. Additionally, they plan their plantings and harvests specifically around the flooding of local rivers. For example, some farmers choose to plant on rising floods and harvest as the flooding diminishes. This techniques is utilized when cultivating rice.

East Africa

East Africa is one of the areas most affected by corporate farming. In western Uganda, there is a farming society called the Banyankore. Their land is part of the Ryampara hill country, between two flat, dry, and less-populated areas of land. On the hills, the average monthly rainfall is 970mm, with two short dry months in June and July and a short, less severe, dry month in January. Seven months out of the year the rainfall is over 1000mm per month. Their primary crop productions are bananas and coffee; they use these as cash crops. Farmers still use substantial areas for millet farming. This is their main food crop. They have intensive home gardens to produce for the families' needs and the outlying farming is used mostly for cash crop production, where coffees and bananas are cultivated using intercropping methods.

Southern Africa

In Southern Africa, conglomerates of farming companies have primarily written-off the lands in Northwest Zambia. The land is mostly made of plateaus, with lower-lying lands in the Kabompo and Zambezi river valleys. Much of the area is in the Congo–Zambezi watershed, the area of land where all the sources of water run into the same river basin. The land has several types of soil. There are areas of yellow clay, with higher concentrations of sand, where higher rainfall causes soil erosion; this makes farming in those areas difficult. There are also areas with fertile red clay; these are rich soils good for cultivation of crops. Farmers in the area grow sorghum, millet, sweet potatoes, pumpkin, and maize. There is marginal maize production, used as a cash crop for the community. The area is usually covered by thick forestation that must be cleared before crops can be cultivated; they do not practice agroforestry. Traditionally, spouses must maintain their own gardens. The produce is shared communally, but the practice leads to intensive home gardens that are maintained by the women of the villages.

SOUTH PACIFIC

Highlander horticulture

The Enga of the Western Highlands Province in New Guinea receive most of their food from growing sweet potatoes Ipomoea batatas which they plant in mulch mounds at elevations up to 2,700 m or higher (Dove and Carpenter 2008). The mounds that the Enga make to plant their crops of potatoes are formed from by piling large amounts of grass taken from fallow, or unplanted, plots then by covering the grass with dirt (Wohlt 2004). The size of the mounds depends on elevation; the higher the elevation; the bigger the mounds will be. Mounds above 2,500 m in altitude can have a height of 0.85 m in height; while crops below 1,500 m are not mounded at all (Dove and Carpenter 2008). The function of the mound is to protect the crops from the frequent frosts that occur at the high altitudes of the Enga. With sweet potatoes having a very long maturation period of nine months, the Enga also invest their time and space on the mounds with planting other crops that have much shorter maturation periods such as peas in case a heavy frost does claim the crop (Wohlt 2004).

The planting of the mounds is done so that the plants which have a higher frost tolerance, such as the Irish potatoes, are planted casually throughout the mound and the low tolerant sweet potatoes are planted in the best position to avoid the frost. Peas, beans, and cabbage which are all highly tolerant to frost will be planted outside the circle of sweet potatoes and lower on the mound placing them closer to the cold temperatures of the ground (Dove and Carpenter 2008). The Enga practice fallow rotation where a garden will in crop for about four years followed by about four years of fallow grassland to let the soil replenish (Wohlt 2004).

Garden size for an average Enga garden is about 0.21 hectare or about 2,100 square meters and can contain a few hundred mounds (Wohlt 2004). Another gardening strategy the Enga have implemented is the use of kin lands that are usually within one to two days walk from the farmers normal planting grounds (Dove and Carpenter 2008). The uses of multiple gardens at differing elevation and the ability access clan lands in different areas for gardening have allowed the Enga to adapt to their environment and survive under harsh conditions.

Island horticulture

For most South Pacific Island cultures the main subsistence techniques are hunting and gathering. Fishing and the gathering of sago, banana, and other tropical foods are the norm with very little organized agriculture. The Tabalu of Kiriwina located in the Trobriand Islands practice a form of agriculture called Kaylu'ebila, a form of garden magic (Malinowski 1965). The main crop for the Tabalu is the yam and there is a definite division of labour according to sex when it comes to gardening. Heavy work is done by the men and it includes clearing the vegetation, caring the yam supports, and planting the yam tubers in the ground (Malinowski 1987). The women aid by weeding the gardens.

Gardening for the Tabalu is a very long and in depth magical process; with special magicians and magical ingredients which have been handed down from family member to family member over time. Garden fields which are called Kwabila are fenced in on all sides to keep out the swine that are breed by the Tabalu. Kwabila are then divided into many smaller plots called baleko, these are the individual gardens that the crops will be planted in (Malinowski 1965).

SOUTH AMERICA

South America consists of modern-day Venezuela, Colombia, Ecuador, Suriname, Brazil, Peru, Bolivia, Paraguay, Uruguay, Argentina, and Chile. South America has historically been a land exploited not only for its natural resources, but also for its indigenous knowledge and labor force. The environmental diversification of South America has been at the foundation of its presence in the global economy as a resource for agriculture, forestry, fishing, hunting, livestock, mining and quarrying.

South America can be seen as cultural regions inhabited by marginal tribes, tropical forest tribes, and the circum-Caribbean tribes each with its own distinct way of agricultural cultivation (Lyon 1974:6). South America's geographic regions are inhabited by regional tribes which include; the Chocó in the Northern Columbian area, The Kayapo in the Eastern Para area, the Chono in the Southern Fuegian area, and the Quechuas in the Western Peruvian area, (Lyon 1974:24). Each of these regions has adapted not only their own cultural identity and agricultural style,

The Eastern region of South America is known as the Para area in what is now Brazil, and has for millennia been home to the tropical forest Kayap? tribe. The Kayap? lived in sedentary villages and were proficient in pottery and loom-weaving, yet they did not domesticate animals or poses knowledge on metallurgy. These Tropical forest Tribes can be characterized through their farming, dugout canoes, woven baskets, loom weaving, and pole and thatch houses (Posey 2002)

In the Para area the Kayap? like most tribes of this region practiced intensive agriculture or clearing cultivation. Beginning their agricultural year with a low water season intensifies fishing. The low water season is then followed by the high water season or harvest season. It is during this harvest season that the Kayap? are able to exercise their leisure time before the cycle ends with (low water levels) and a return to intense fishing. Each changing season commences ceremonies for the Kayap? that are directly tied to agriculture, hunting, or fishing. Unlike the Chocó the Kayap? used an agricultural method known as the slash-burn method (shifting agriculture). The Kayap? cut the forest in April to September (dry season) and time their burns just before the raining season. The Kayap? used circular plots for agricultural cultivation consisting of five rings (characterized as cultivation zones). The first circle or the inner circle was used for taro and sweet potatoes that thrive in the hotter soil found in the center of the plot. The second circle cultivated maize, manioc, and rice an area that needed various ash enrichment treatments and would experience short term growth. The third zone was an area of rich soil and best served mixed crops including the banana, urucu, papaya. The fourth zones consists of shade loving plants and are for a medical purpose, yet evidence of beans and yams have been also found here. The fifth zone or the outer ring was left as a protective zone that included trap animals protective insects and birds (Posey 2002). This form of agriculture requires not only intense physical labor but also requires a knowledge of not only the land, but various types of ground cover, shades and temperatures of local soils, as well as cloud formations to time careful burning. When the Kayap? manage their agricultural plots they must work with a variety of interacting factors including the background soil fertility, the heterogeneous quality of ash and its distribution, crop nutrient requirements, cropping cycles, management requirements, and pest and disease control clearly illustrating the common misconception that this form of agriculture is primitive and ineffiecnet (Posey 2002).

It has often been thought that the slash-burn plots are abandoned after one or two years because of un productive soil, but this is a common misconception. The Kayap? revisit abandoned fields because plants can offer direct and indirect benefits. One direct benefit would be the ability to eat that which has been produced and an indirect benefit would be that open fields allow attract game for hunting and can produce long after they have been tended (Posey 2002).

The Southern region of South America is known as the Fuegian area and is occupied by the Chono, Alacaluf and Yahgan. These Marginal tribes differed greatly from the other regions in that they were expert in making bark or plank canoes and domesticating dogs, hunting, fishing and gathering. Nomads with simple socio-religious patterns, yet completely lacked the technology of pottery, loom-weaving, metallurgy and even agriculture. Since there was a lack of agriculture in this region the Chono ate native berries, roots, fruits, and wild celery. The source of nutrition for the Chono, Yahgans, and Alcaaluf thus mainly consisted of sea food such as; whales, seal porpoises, guanacos, and otters. The Southern tribes of South America distinguished themselves not in having a rare form of agriculture like the North's slush-mush method or the having intensive agriculture like the East's sophisticated slash-burn method, yet were able to distinguished themselves in being absent of this trait.

TROPICAL HORTICULTURE

Tropical horticulture is a branch of horticulture that studies and cultivates plants in the tropics, i.e., the equatorial regions of the world. The field is sometimes known by the portmanteau "TropHort". Tropical horticulture includes plants such as perennial woody plants (arboriculture), ornamentals (floriculture), vegetables (olericulture), and fruits (pomology) including grapes (viticulture). The origin of many of these crops is not in the tropics but in temperate zones. Their adoption to tropical climatic conditions is an objective of breeding. Many important crops, however, are indigenous to the tropics. The latter embrace perennial crops such as oil palm, vegetables including okra, field crops such as rice and sugarcane, and particularly fruits including pineapple, banana, papaya, and mango.

Since the tropics represent 36 percent of the earth's surface and 20 percent of its land surface, the potential of tropical horticulture is huge. In contrast to temperate regions, environmental conditions in the tropics are defined less by seasonal temperature fluctuations and more by seasonality of precipitation. Thus the climate in the greater part of the tropics is characterized by distinct wet and dry seasons, although such variation is reduced in locations closer to the equator (±5° latitude). Temperature conditions in the tropics are affected by elevation, in which contrasting warmer and colder climate areas in the tropics can be differentiated, and highland areas in the tropics can consequently be more favourable for production of temperate plant species than are lowland areas.

HORTICULTURAL FLORA

A horticultural flora, also known as a garden flora, is a plant identification aid structured in the same way as a native plants flora. It serves the same purpose: to facilitate plant identification; however, it only includes plants that are under cultivation as ornamental plants growing within the prescribed climate zone or region. Traditionally published in book form, often in several volumes, such floras are increasingly likely to be produced as websites or CD ROMs.

Uses

Written by a professional plant taxonomist or plantsperson, a horticultural flora assists clarification of scientific and common names, the identification of plant characteristics that occur in cultivated plants that are additional to those in wild counterparts, and descriptions of cultivars.

Although horticultural floras may include a range of food plants, their emphasis is generally on ornamental plants and so these floras are sometimes referred to as "garden floras". Increasingly they provide data for sustainable landscaping, such as:

- Drought tolerance and irrigation needs
- Food sources of native and migratory bird and butterfly species
- Companion planting
- Invasive species notation
- Local habitat restoration aspects.

Horticultural oil

Horticultural oils or narrow range oils are lightweight oils, either petroleum or vegetable based. They are used in both horticulture and agriculture, where they are applied as a dilute spray on plant surfaces to control insects and mites. They are also sometimes included in tank mixes as a surfactant.

The oils provide control by smothering the target pests, and are only effective if applied directly to the pest, and provide no residual controls. Oils are generally considered suitable for 'organic pest control', with most oils permitted under the U.S. National Organic Program.

Types of oils

Dormant oil: An oil used on woody plants during the dormant season. This term originally referred to heavier weight, less well-refined oils that were unsafe to use on plants after they broke dormancy. However, these older oils have been replaced with more refined, light-weight oils that have potential application to plant foliage. Dormant oil now refers to the time of application rather than to any characteristic type of oil.

Horticultural oils: An oil used to control a pest on plants.

Mineral oil: A petroleum-derived oil (as opposed to vegetable oils).

Narrow-range oil: A highly refined oil that has a narrow range of distillation. Narrow-range oils fall in the superior oil classification. The terms may be used nearly interchangeably.

Spray oil: An oil designed to be mixed with water and applied to plants as a spray for pest control.

Summer oil: An oil used on plants when foliage is present (also called foliar oils). As with dormant oil, the term now refers to the time an application is made rather than to the properties of the oil.

Supreme oil: A term used to categorize highly refined oils that distill at slightly higher temperatures and over a wider range than the narrow-range oils. Most supreme oils meet the characteristics of a superior oil.

Superior oil: A term originated by P.J. Chapman in 1947 to categorize summer-use oils that met certain specifications. This included a high proportion of paraffinic hydrocarbons and purification that allowed year-round use without phytotoxicity. Since then, further developments have resulted in oils that distill over a narrow temperature range. Most superior oils are now better referred to as narrow-range oils.

Vegetable oil: An oil derived from the seeds of some oil seed crop (e.g., soybeans, canola, cottonseed).

Horticulture Farming in India

India has diverse agro-climatic conditions, all sorts of fresh fruits, vegetables, and medicinal plants can be grown in different regions of the nation. In addition, There is a growing health consciousness among the Indian population.

Since the majority of the population is vegetarian in India. There is a strong demand for fruits and vegetables. However, When compared to the country's current demand for horticulture products, the country's production is far lower. Hence, there is huge scope for farmers and businesses in the Horticulture sector in India.

Types of Horticulture Farming in India

There are numerous types of Horticulture Farming in India. However, the most significant ones are as follows.

Arboriculture

How plants grow and adapt to their environment is studied in arboriculture. In arboriculture, Individual trees, shrubs, vines, and other perennial woody plants are studied. In simple words, it generally relates to maintaining individual woody plants and trees for long-term landscape and amenity purposes.

Turf Management

Turf Management is one of the important types of horticulture farming in India. It describes the work to manage the sports pitches. For example, management of cricket pitch, maintenance of golf pitch etc. comes under Turf Management.

Floriculture

Floriculture is the process of farming flowers. It is a branch of horticulture associated with flower and ornamental crops for use in gardens and floristry. Floriculturists spend most of their time cultivating new varieties of flowers through plant breeding.

In simple words, it is the growing, maintenance, and marketing of ornamental flowers. However, some florists also do the extraction of essential oils, edible pigments.

To get a detailed overview of Floriculture, you can visit – An Overview of Floriculture

Landscape Horticulture

In landscape gardening, attractive plants are planted in such a manner as to create a scenic appearance. In simple words, Landscape gardening is an artistic branch of Horticulture.

Landscape Horticulture is a mixture of science and art. It is a vital and emerging sector considering the scope of horticulture farming in India.

Pomology

Pomology is the study of fruit, specifically the science of growing fruit and nuts. Fruits are always on-high demand. Therefore, Pomology has a significant role in horticulture farming of India. Among a pomologist's responsibilities is the creation of new cultivars of fruits and vegetables. Fruit and nut cultivars are continuously being modified to enhance disease resistance and other characteristics. Moreover, Pomologists also investigate fertiliser and pruning strategies for growing healthy trees.

Postharvest Physiology

Postharvest Physiology is used to determine the best storage and transportation conditions to optimise shelf life after harvest.

As the name suggests, Postharvest Physiology focuses primarily on technology after harvesting. In Postharvest Physiology, we also get to study optimal storage and transportation conditions to increase the shelf life.

Commercial Horticulture Farming

As we all know, People make money from commercial horticulture by cultivating plants for other people.

It requires a thorough understanding of the market, including what customers want, what their likes and preferences are, and what is readily accessible and what is in short supply.

Let us now read about the Benefits of horticulture farming in India.

Benefits of Horticulture Farming in India

Horticulture benefits are not limited to the environment only. Rather, Horticulture farming can be a profitable business in the long run. The reason is it provides several benefits. Let us look at some of the most important ones.

Horticulture is one of India's most important development drivers. It can be more lucrative than agriculture in several cases (food grains mainly).

Horticulture offers a variety of professions in the primary, secondary, and tertiary sectors, as well as in the government.

Horticulture also allows people to consume a diversified and balanced food, which is important for a healthy lifestyle.

Exporters of scheduled horticulture products having APEDA Registration can expand their business worldwide and earn profits. Moreover, APEDA Registration allows exporters and farmers to have financial aid and attend various training programs to expand the business.

2

GROWTH REGULATION OF FRUIT CROPS

The plant hormones are extremely important agent in the integration of developmental activities. Environmental factors often exert inductive effects by evoking changes in hormones in metabolism and distribution within the plant. Apart from it, they also regulate expression of intrinsic genetic potential of plants. Control of genetic expression has been demonstrated for the phytohormones at both transcriptional and translational levels. Also, hormones receptors and binding proteins have been identified on membrane surface that are specific for some hormones.The use of growth regulators has become an important component of agro-technical procedures for most of the cultivated plants and especially for fruit plants. So far in fruit crops, excessive fruit drop can be controlled by the exogenous application of plant growth regulators. The auxin and gibberellins are widely used to control the fruit drop and to improve the quality of fruit. Ontogenic development from fruit set to fruit ripening and final reach to customer, several agents are responsible for elimination of some fruits from fruit set to final maturity. In this review, we focus on the major functions of plant growth regulators in fruit production

To evaluate the effects of plant growth regulators on quality and quantity of pomegranate fruits, 2,4-dichlorophenoxyacetic acid (2,4-D), naphthalene acetic acid (NAA) and naphthalene acetamide (NAD) were applied to 'Shisheh Cup' pomegranate trees in 1999 and 2000. In the first experiment 2,4-D (15, 30, 45 mg/L), NAA (25, 50, 100 mg/L) and NAD (50, 100, 200 mg/L) were applied to the uniform branch units of pomegranate trees at full bloom, 45 and 90 days after full bloom. In the second experiment 2,4-D (15, 30, 45 mg/L) were applied at full bloom, 30, 60 and 90 days after full bloom to the uniform branch units of pomegranate trees. In both experiments treatments were applied with a hand sprayer until the drip point and each treatment was applied to 3 trees and 4 uniform branch units per tree. Fruit samples of treated branches were harvested at maturity stage and qualitative and quantitative characters of fruits were evaluated. In both experiments plant growth regulators significantly increased fruit L/D ratio, fruit weight, fruit volume, (expectedly NAD) rind weight, aril weight, and leaf area index. For the best effects of plant growth regulators on qualitative and quantitative characters in pomegranate fruits, it is recommended to apply 2,4-D (30 mg/L), NAA (50 mg/L) and NAD (100 mg/L)..

PLANT GROWTH REGULATORS

We all know that plants need light, water, oxygen and nutrition to grow and develop. All these qualify as extrinsic factors. While extrinsic factors are important, did you know that plant growth depends on intrinsic factors too? They can be intracellular genes or intercellular chemicals. These chemicals are called Plant Growth Regulators. Let's learn about them in more detail below.

Plant Growth Regulators are defined as small, simple chemicals produced naturally by plants to regulate their growth and development.

Characteristics

Plant Growth Regulators can be of a diverse chemical composition such as gases (ethylene), terpenes (gibberellic acid) or carotenoid derivates (abscisic acid). They are also referred to as plant growth substances, phytohormones or plant hormones. Based on their action, they are broadly classified as follows:

- Plant Growth Promoters – They promote cell division, cell enlargement, flowering, fruiting and seed formation. Examples are auxins, gibberellins and cytokinins.
- Plant Growth Inhibitors – These chemicals inhibit growth and promote dormancy and abscission in plants. An example is an abscisic acid.

Note: Ethylene can be a promoter or an inhibitor, but is largely a Plant Growth Inhibitor. All plant growth regulators were discovered accidentally. Let's take a detailed look at each regulator and learn about it more closely:

AUXINS

Discovery

Auxins were the first growth hormone to be discovered. They were discovered due to the observations of Charles Darwin and his son, Francis Darwin. The Darwins observed that the coleoptile (protective sheath) in canary grass grows and bends towards the source of light. This phenomenon is 'phototropism'. In addition, their experiments showed that the coleoptile tip was the site responsible for the bending. Finally, this led to the isolation of the first auxin by F. W. Went from the coleoptile tip of oat seedlings.

Auxins are a class of plant hormones (or plant-growth regulators) with some morphogen-like characteristics. Auxins play a cardinal role in coordination of many growth and behavioral processes in plant life cycles and are essential for plant body development. The Dutch biologist Frits Warmolt Went first described auxins and their role in plant growth in the 1920s.Kenneth V. Thimann (1904-1997) became the first to isolate one of these phytohormones and to determine its chemical structure as indole-3-acetic acid (IAA). Went and Thimann co-authored a book on plant hormones, Phytohormones, in 1937.

Overview

Auxins were the first of the major plant hormones to be discovered. They derive their name from the Greek word áõîåéí (auxein – "to grow/increase"). Auxin (namely IAA) is present in all parts of a plant, although in very different concentrations. The concentration in each position is crucial developmental information, so it is subject to tight regulation through both metabolism and transport. The result is the auxin creates "patterns" of auxin concentration maxima and minima in the plant body, which in turn guide further development of respective cells, and ultimately of the plant as a whole. The (dynamic and environment responsive) pattern of auxin distribution within the plant is a key factor for plant growth, its reaction to its environment, and specifically for development of plant organs (such as leaves or flowers). It is achieved through very complex and well-coordinated active transport of auxin molecules from cell to cell throughout the plant body — by the so-called polar auxin transport. Thus, a plant can (as a whole) react to external conditions and adjust to them, without requiring a nervous system. Auxins typically act in concert with, or in opposition to, other plant hormones. For example, the ratio of auxin to cytokinin in certain plant tissues determines initiation of root versus shoot buds. On the molecular level, all auxins are compounds with an aromatic ring and a carboxylic acid group. The most important member of the auxin family is indole-

3-acetic acid (IAA), which generates the majority of auxin effects in intact plants, and is the most potent native auxin. And as native auxin, its stability is controlled in many ways in plants, from synthesis, through possible conjugation to degradation of its molecules, always according to the requirements of the situation.

- Five naturally occurring (endogenous) auxins in plants include indole-3-acetic acid, 4-chloroindole-3-acetic acid, phenylacetic acid, indole-3-butyric acid, and indole-3-propionic acid. However, most of the knowledge described so far in auxin biology and as described in the article below, apply basically to IAA; the other three endogenous auxins seems to have rather marginal importance for intact plants in natural environments. Alongside endogenous auxins, scientists and manufacturers have developed many synthetic compounds with auxinic activity.
- Synthetic auxin analogs include 1-naphthaleneacetic acid, 2,4-dichloro phenoxyacetic acid (2,4-D), and many others.

Some synthetic auxins, such as 2,4-D and 2,4,5-trichlorophenoxyacetic acid (2,4,5-T), are used also as herbicides. Broad-leaf plants (dicots), such as dandelions, are much more susceptible to auxins than narrow-leaf plants (monocots) such as grasses and cereal crops, so these synthetic auxins are valuable as synthetic herbicides.

DISCOVERY OF AUXIN

Charles Darwin

In 1881, Charles Darwin and his son Francis performed experiments on coleoptiles, the sheaths enclosing young leaves in germinating grass seedlings. The experiment exposed the coleoptile to light from a unidirectional source, and observed that they bend towards the light. By covering various parts of the coleoptiles with a light-impermeable opaque cap, the Darwins discovered that light is detected by the coleoptile tip, but that bending occurs in the hypocotyl. However the seedlings showed no signs of development towards light if the tip was covered with an opaque cap, or if the tip was removed. The Darwins concluded that the tip of the coleoptile was responsible for sensing light, and proposed that a messenger is transmitted in a downward direction from the tip of the coleoptile, causing it to bend.

Peter Boysen-Jensen

In 1913, Danish scientist Peter Boysen-Jensen demonstrated that the signal was not transfixed but mobile. He separated the tip from the remainder of the coleoptile by a cube of gelatine which prevented cellular contact but allowed chemicals to pass through. The seedlings responded normally bending towards the light. However, when the tip was separated by an impermeable substance, there was no curvature of the stem.

Frits Went

In 1928, the Dutch botanist Frits Warmolt Went showed that a chemical messenger diffuses from coleoptile tips. Went's experiment identified how a growth promoting chemical causes a coleoptile to grow towards the light. Went cut the tips of the coleoptiles and placed them in the dark, putting a few tips on agar blocks that he predicted would absorb the growth-promoting chemical. On control coleoptiles, he placed a block that lacked the chemical. On others, he placed blocks containing the chemical, either centered on top of the coleoptile to distribute the chemical evenly or offset to increase the concentration on one side. When the growth-promoting chemical was distributed evenly the coleoptile grew straight. If the chemical was distributed unevenly, the coleoptile curved away from the side with the cube, as if growing towards the light, even though it was grown in the dark. Went later proposed that the messenger substance is a growth-promoting hormone, which he named auxin, that becomes asymmetrically distributed in the bending region. Went concluded that auxin is at a higher concentration on the shaded side, promoting cell elongation, which results in coleoptiles bending towards the light.

Hormonal activity

Auxins help development at all levels in plants, from the cellular level, through organs, and ultimately to the whole plant.

Molecular mechanisms

When a plant cell comes into contact with auxin, it causes dramatic changes in gene expression, with many genes up- or down-regulated. The precise mechanisms by which this occurs are still an area of active research, but there is now a general consensus on at least two auxin signalling pathways.

Perception

The most well-characterized auxin receptors are the TIR1/ AFB family of F-box proteins. F-box proteins target other proteins for degradation via the ubiquitin degradation pathway. When TIR1/ ABF proteins bind to auxin, the auxin acts as a 'molecular glue' that allows these proteins to then bind to their targets (see below).

Another auxin-binding protein, APB1 is now often regarded as an auxin receptor (at the apoplast), but it is generally considered to have a much more minor role than the TIR1/ABF signaling pathway, and much less is known about APB1 signaling.The auxin signal cascade: In the absence of auxin, Aux/IAA bind to and suppress the transcriptional activity of ARFs. When auxin is present it forms a 'molecular glue' between TIR1 and Aux/IAAs, which leads to the degradation of these repressors. ARFs are then free to bind to DNA and to cause changes in transcription.

Auxin response factors (ARFs) are a large group of transcription factors that act in auxin signaling. In the absence of auxin, ARFs bind to a class of repressors known as Aux/IAAs. Aux/IAA suppress the ability of ARFs to enhance gene transcription. Additionally, the binding of Aux/IAA to ARFs brings Aux/IAA into contact with the promoters of auxin-regulated genes. When at these promoters, Aux/IAA repress the expression of these genes through recruiting other factors to make modifications to the DNA structure.

The binding of auxin to TIR1/AFBs allows them to bind to Aux/IAAs. When bound by TIR1/AFBs, Aux/IAAs are marked for degradation. The degradation of Aux/IAA frees ARF proteins, which are then able to activate or repress genes at whose promoters they are bound.

The large number of Aux/IAA and ARF binding pairs possible, and there different distributions between cell types and across developmental age are thought to account for the astonishingly diverse responses that auxin produces.

In June 2018 it was demonstrated that plant tissues can respond to auxin in a TIR1-dependent manner extremely quickly (probably too quickly to be explained by changes in gene expression). This has led some scientists to suggest that there is an as yet unidentified TIR1-dependent auxin-signalling pathway that differs from the well-known transcriptional response.

On a cellular level

On the cellular level, auxin is essential for cell growth, affecting both cell division and cellular expansion. Auxin concentration level, together with other local factors, contributes to cell differentiation and specification of the cell fate. Depending on the specific tissue, auxin may promote axial elongation (as in shoots), lateral expansion (as in root swelling), or isodiametric expansion (as in fruit growth). In some cases (coleoptile growth), auxin-promoted cellular expansion occurs in the absence of cell division. In other cases, auxin-promoted cell division and cell expansion may be closely sequenced within the same tissue (root initiation, fruit growth). In a living plant, auxins and other plant hormones nearly always appear to interact to determine patterns of plant development.

Organ Patterns

Growth and division of plant cells together result in the growth of tissue, and specific tissue growth contributes to the development of plant organs. Growth of cells contributes to the plant's size, unevenly localized growth produces bending, turning and directionalization of organs- for example, stems turning toward light sources (phototropism), roots growing in response to gravity (gravitropism), and other tropisms originated because cells on one side grow faster than the cells on the other side of the organ. So, precise control of auxin distribution between different cells has paramount importance to the resulting form of plant growth and organization.

Auxin transport and the Uneven Distribution of auxin

To cause growth in the required domains, auxins must of necessity be active preferentially in them. Local auxin maxima can be formed by active biosynthesis in certain cells of tissues, for example via tryptophan-dependent pathways, but auxins are not synthesized in all cells (even if cells retain the potential ability to do so, only under specific conditions will auxin synthesis be activated in them). For that purpose, auxins have to be not only translocated toward those sites where they are needed but also they must have an established mechanism to detect those sites. For that purpose, auxins have to be translocated toward those sites where they are needed. Translocation is driven throughout the plant body, primarily from peaks of shoots to peaks of roots (from up to down).

For long distances, relocation occurs via the stream of fluid in phloem vessels, but, for short-distance transport, a unique system of coordinated polar transport directly from cell to cell is exploited. This short-distance, active transport exhibits some morphogenetic properties. This process, polar auxin transport, is directional, very strictly regulated, and based in uneven distribution of auxin efflux carriers on the plasma membrane, which send auxins in the proper direction. While PIN-FORMED (PIN) proteins are vital in transporting auxin in a polar manner, the family of AUXIN1/LIKE-AUX1 (AUX/LAX) genes encodes for non-polar auxin influx carriers. The regulation of PIN protein localisation in a cell determines the direction of auxin transport from cell, and concentrated effort of many cells creates peaks of auxin, or auxin maxima (regions having cells with higher auxin – a maximum). Proper and timely auxin maxima within developing roots and shoots are necessary to organise the development of the organ. Surrounding auxin maxima are cells with low auxin troughs, or auxin minima. For example, in the Arabidopsis fruit, auxin minima have been shown to be important for its tissue development.

Organization of the plant

As auxins contribute to organ shaping, they are also fundamentally required for proper development of the plant itself. Without hormonal regulation and organization, plants would be merely proliferating heaps of similar cells. Auxin employment begins in the embryo of the plant, where the directional distribution of auxin ushers in subsequent growth and development of primary growth poles, then forms buds of future organs. Next, it helps to coordinate correct development of the arising organs, like roots, cotyledons, and leaves and mediates long-distance signals between them, conducive therefore to the general designof the plant. Throughout the plant's life, auxin helps the plant maintain the polarity of growth, and actually "recognize" where it has its branches (or any organ) connected. An important principle of plant organization based upon auxin distribution is apical dominance, which means the auxin produced by the apical bud (or growing tip) diffuses (and is transported) downwards and inhibits the development of ulterior lateral bud growth, which would otherwise compete with the apical tip for light and nutrients. Removing the apical tip and its suppressively acting auxin allows the lower dormant lateral buds to develop, and the buds between the leaf stalk and stem produce new shoots which compete to become the lead growth. The process is actually quite complex because auxin transported downwards from the lead shoot tip has to interact with several other plant hormones (such as strigolactones or cytokinins) in the process on various positions along the growth axis in plant body to achieve this phenomenon. This plant behavior is used in pruning by horticulturists.

Finally, the sum of auxin arriving from stems to roots influences the degree of root growth. If shoot tips are removed, the plant does not react just by the outgrowth of lateral buds — which are supposed to replace to original lead. It also follows that smaller amount of auxin arriving at the roots results in slower growth of roots and the nutrients are subsequently in higher degree invested in the upper part of the plant, which hence starts to grow faster.

Effects

Crown galls are caused by Agrobacterium tumefaciensbacteria; they produce and excrete auxin and cytokinin, which interfere with normal cell division and cause tumors. Auxin participates in phototropism, geotropism, hydrotropism and other developmental changes. The uneven distribution of auxin, due to environmental cues, such as unidirectional light or gravity force, results in uneven plant tissue growth, and generally, auxin governs the form and shape of the plant body, direction and strength of growth of all organs, and their mutual interaction.

Auxin stimulates cell elongation by stimulating wall-loosening factors, such as elastins, to loosen cell walls. The effect is stronger if gibberellins are also present. Auxin also stimulates cell division if cytokinins are present. When auxin and cytokinin are applied to callus, rooting can be generated if the auxin concentration is higher than cytokinin concentration. Xylem tissues can be generated when the auxin concentration is equal to the cytokinins. Auxin also induces sugar and mineral accumulation at the site of application.

WOUND RESPONSE

Auxin induces the formation and organization of phloem and xylem. When the plant is wounded, the auxin may induce the cell differentiation and regeneration of the vascular tissues.

Root Growth and Development

Auxins promote root initiation. Auxin induces both growth of pre-existing roots and adventitious root formation, i.e., branching of the roots. As more native auxin is transported down the stem to the roots, the overall development of the roots is stimulated. If the source of auxin is removed, such as by trimming the tips of stems, the roots are less stimulated accordingly, and growth of stem is supported instead. In horticulture, auxins, especially NAA and IBA, are commonly applied to stimulate root initiation when rooting cuttings of plants. However, high concentrations of auxin inhibit root elongation and instead enhance adventitious root formation. Removal of the root tip can lead to inhibition of secondary root formation.

Apical dominance

Auxin induces shoot apical dominance; the axillary buds are inhibited by auxin, as a high concentration of auxin directly stimulates ethylene synthesis in axillary buds, causing inhibition of their growth and potentiation of apical dominance. When the apex of the plant is removed, the inhibitory effect is removed and the growth of lateral buds is enhanced. Auxin is sent to the part of the plant facing away from the light, where it promotes cell elongation, thus causing the plant to bend towards the light.

Fruit growth and development

Auxin is required for fruit growth and development and delays fruit senescence. When seeds are removed from strawberries, fruit growth is stopped; exogenous auxin stimulates the growth in fruits with seeds removed. For fruit with unfertilized seeds, exogenous auxin results in parthenocarpy (“virgin-fruit” growth). Fruits form abnormal morphologies when auxin transport is disturbed. In Arabidopsis fruits, auxin controls the release of seeds from the fruit (pod). The valve margins are a specialised tissue in pods that regulates when pod will open (dehiscence). Auxin must be removed from the valve margin cells to allow the valve margins to form. This process requires modification of the auxin transporters (PIN proteins).

Flowering

Auxin plays also a minor role in the initiation of flowering and development of reproductive organs. In low concentrations, it can delay the senescence of flowers. A number of plant mutants have been described that affect flowering and have deficiencies in either auxin synthesis or transport. In maize, one example is bif2 barren inflorescence2.

Ethylene biosynthesis

In low concentrations, auxin can inhibit ethylene formation and transport of precursor in plants; however, high concentrations can induce the synthesis of ethylene. Therefore, the high concentration can induce femaleness of flowers in some species.

Auxin inhibits abscission prior to the formation of the abscission layer, and thus inhibits senescence of leaves.

SYNTHETIC AUXINS

In the course of research on auxin biology, many compounds with noticeable auxin activity were synthesized. Many of them had been found to have economical potential for man-controlled growth and development of plants in agronomy. Synthetic auxins include the following compounds:

Auxins are toxic to plants in large concentrations; they are most toxic to dicots and less so to monocots. Because of this property, synthetic auxin herbicides, including 2,4-D and 2,4,5-T, have been developed and used for weed control.

However, some exogenously synthesized auxins, especially 1-naphthaleneacetic acid (NAA) and indole-3-butyric acid (IBA), are also commonly applied to stimulate root growth when taking cuttings of plants or for different agricultural purposes such as the prevention of fruit drop in orchards. Used in high doses, auxin stimulates the production of ethylene. Excess ethylene (also native plant hormone) can inhibit elongation growth, cause leaves to fall (abscission), and even kill the plant. Some synthetic auxins, such as 2,4-D and 2,4,5-trichlorophenoxyacetic acid (2,4,5-T) were marketed also as herbicides. Dicots, such as dandelions, are much more susceptible to auxins than monocots, such as grasses and cereal crops. So, these synthetic auxins are valuable as synthetic herbicides. 2,4-D was the first widely used herbicide, and it is still so.2,4-D was first commercialized by the Sherwin-Williams company and saw use in the late 1940s. It is easy and inexpensive to manufacture.

Gibberellins

The first inroads into the understanding of GAs were developments from the plant pathology field, with studies on the bakanae, or "foolish seedling" disease in rice. Foolish seedling disease causes a strong elongation of rice stems and leaves and eventually causes them to topple over. In 1926, Japanese scientist Eiichi Kurosawa identified that foolish seedling disease was caused by the fungusGibberella fujikuroi. Later work at the University of Tokyo (notable from Yabuta, Sumiki and Hayashi) showed that a substance produced by this fungus triggered the symptoms of foolish seedling disease and they named this substance "gibberellin".

The increased communication between Japan and the west following World War II enhanced the interest in gibberellin in the United Kingdom (UK) and the United States (US). Workers at Imperial Chemical Industries in the UK and the Department of Agriculture in the US both independently isolated gibberellic acid (with the Americans originally referring to the chemical as "gibberellin-X", before adopting the British name and the chemical is known as gibberellin A3 or GA3 in Japan) Knowledge of gibberellins spread around the world as the potential for its use on various commercially important plants became more obvious. For example, research that started at the University of California, Davis in the mid-1960s led to its commercial use on Thompson seedless table grapes throughout California by 1962. A known gibberellin biosynthesis inhibitor is paclobutrazol (PBZ), which in turn inhibits growth and induces early fruitset as well as seedset. A chronic food shortage was feared during the rapid climb in world population in the 1960s. This was averted with the development of a high-yielding variety of rice. This variety of semi-dwarf rice is called IR8, and it has a short height because of a mutation in the sd1 gene. Sd1 encodes GA20ox, so a mutant sd1 is expected to exhibit a short height that is consistent with GA deficiency.

Chemistry

All known gibberellins are diterpenoid acids that are synthesized by the terpenoid pathway in plastids and then modified in the endoplasmic reticulum and cytosol until they reach their biologically-active form. All gibberellins are derived via the ent-gibberellane skeleton, but are synthesised via ent-kaurene. The gibberellins are named GA1 through GAn in order of discovery. Gibberellic acid, which was the first gibberellin to be structurally characterized, is GA3.

As of 2003, there were 126 GAs Identified from Plants, fungi, and Bacteria.

Gibberellins are tetracyclic diterpene acids. There are two classes based on the presence of either 19 or 20 carbons. The 19-carbon gibberellins, such as gibberellic acid, have lost carbon 20 and, in place, possess a five-member lactone bridge that links carbons 4 and 10. The 19-carbon forms are, in general, the biologically active forms of gibberellins. Hydroxylation also has a great effect on the biological activity of the gibberellin. In general, the most biologically active compounds are dihydroxylated gibberellins, which possess hydroxyl groups on both carbon 3 and carbon 13. Gibberellic acid is a dihydroxylated gibberellin.

Bioactive GAs

The bioactive GAs are GA1, GA3, GA4, and GA7. There are three common structural traits between these GAs: 1) a hydroxyl group on C-3â, 2) a carboxyl group on C-6, and 3) a lactone between C-4 and C-10. The 3â-hydroxyl group can be exchanged for other functional groups at C-2 and/or C-3 positions. GA5 and GA6 are examples of bioactive GAs that do not have a hydroxyl group on C-3â. The presence of GA1 in various plant species suggests that it is a common bioactive GA.

BIOLOGICAL FUNCTION

1. Shows a plant lacking gibberellins and has an internode length of "0" as well as it is a dwarf plant. 2. Shows your average plant with a moderate amount of gibberellins and an average internode length. 3.Shows a plant with a large amount of gibberellins and so has a much longer internode length because gibberellins promotes cell division in the stem.

Gibberellins are involved in the natural process of breaking dormancy and other aspects of germination. Before the photosynthetic apparatus develops sufficiently in the early stages of germination, the stored energy reserves of starch nourish the seedling. Usually in germination, the breakdown of starch to glucose in the endosperm begins shortly after the seed is exposed to water. Gibberellins in the seed embryo are believed to signal starch hydrolysis through inducing the synthesis of the enzyme á-amylase in the aleurone cells. In the model for gibberellin-induced production of á-amylase, it is demonstrated that gibberellins (denoted by GA) produced in the scutellum diffuse to the aleurone cells, where they stimulate the secretion á-amylase. á-Amylase then hydrolyses starch, which is abundant in many seeds, into glucose that can be used in cellular respiration to produce energy for the seed embryo. Studies of this process have indicated gibberellins cause higher levels of transcription of the gene coding for the á-amylase enzyme, to stimulate the synthesis of á-amylase. Gibberellins are produced in greater mass when the plant is exposed to cold temperatures. They stimulate cell elongation, breaking and budding, seedless fruits, and seed germination. They do the last by breaking the seed's dormancy and acting as a chemical messenger. Its hormone binds to a receptor, and Ca activates the protein calmodulin, and the complex binds to DNA, producing an enzyme to stimulate growth in the embryo.

METABOLISM

Biosynthesis

GAs are usually synthesized from the methylerythritol phosphate (MEP) pathway in higher plants. In this pathway, bioactive GA is produced from trans-geranylgeranyl diphosphate (GGDP). In the MEP pathway, three classes of enzymes are used to yield GA from GGDP: 1) terpene synthases (TPSs), 2) cytochrome P450 monooxygenases (P450s), and 3) 2-oxoglutarate–dependent dioxygenases (2ODDs). There are 8 steps in the methylerythritol phosphate pathway: - 1) GGDP is converted to ent-copalyl diphosphate (ent-CPD) by ent-copalyl diphosphate synthase - 2) etn-CDP is converted to ent-kaurene by ent-kaurene synthase - 3) ent-kaurene is converted to ent-kaurenol by ent-kaurene oxidase (KO) - 4) ent-kaurenol is converted to ent-kaurenal by KO - 5) ent-kaurenal is converted to ent-kaurenoic acid by KO - 6) ent-kaurenoic acid is converted to ent-7a-hydroxykaurenoic acid by ent-kaurene acid oxidase (KAO) - 7) ent-7a-hydroxykaurenoic acid is converted to GA12-aldehyde by KAO - 8) GA12-aldehyde is converted to GA12 by KAO. GA12 is processed to the bioactive GA4 by oxidations on C-20 and C-3, which is accomplished by 2 soluble ODDs: GA 20-oxidase and GA 3-oxidase. One or two genes encode the enzymes responsible for the first steps of GA biosynthesis in Arabidopsis and rice. The null alleles of the genes encoding CPS, KS, and KO result in GA-deficient Arabidopsis dwarves. Multigene families encode the 2ODDs that catalyze the formation of GA12 to bioactive GA4. AtGA3ox1 and AtGA3ox2, two of the four genes that encode GA3ox in Arabidopsis, affect vegetative development. Environmental stimuli regulate AtGA3ox1 and AtGA3ox2 activity during seed germination. In Arabidopsis, GA20ox overexpression leads to an increase in GA concentration.

Sites of biosynthesis

Most bioactive GAs are located in actively growing organs on plants. Both GA20ox and GA3ox genes (genes coding for GA 20-oxidase and GA 3-oxidase) and the SLENDER1 gene (a GA signal transduction gene) are found in growing organs on rice, which suggests bioactive GA synthesis occurs at their site of action in growing organs in plants. During flower development, the tapetum of anthers is believed to be a primary site of GA biosynthesis.

Differences Between Biosynthesis in Fungi and lower Plants

Arabidopsis, a plant, and Gibberella fujikuroi, a fungus, possess different GA pathways and enzymes. P450s in fungi perform functions analogous to the functions of KAOs in plants. The function of CPS and KS in plants is performed by a single enzyme, CPS/KS, in fungi. In fungi, the GA biosynthesis genes are found on one chromosome, but in plants, they are found randomly on multiple chromosomes. Plants produce low amount of GA3, therefore the GA3 is produced for industrial purposes by microorganisms. Industrially the gibberellic acid can be produced by submerged fermentation, but this process presents low yield with high production costs and hence higher sale value, nevertheless other alternative process to reduce costs of the GA3 production is Solid-State Fermentation (SSF) that allows the use of agro-industrial residues.

Catabolism

Several mechanisms for inactivating GAs have been identified. 2â-hydroxylation deactivates GA, and is catalyzed by GA2-oxidases (GA2oxs). Some GA2oxs use C19-GAs as substrates, and other GA2oxs use C20-GAs. Cytochrome P450 mono-oxygenase, encoded by elongated uppermost internode (eui), converts GAs into 16á,17-epoxides. Rice eui mutants amass bioactive GAs at high levels, which suggests cytochrome P450 mono-oxygenase is a main enzyme responsible for deactivation GA in rice. The Gamt1 and gamt2 genes encode enzymes that methylate the C-6 carboxyl group of GAs. In a gamt1 and gamt2 mutant, concentrations of GA is developing seeds is increased.

Homeostasis

Feedback and feedforward regulation maintains the levels of bioactive GAs in plants. Levels of AtGA20ox1 and AtGA3ox1 expression are increased in a GA deficient environment, and decreased after the addition of bioactive GAs, Conversely, expression of At GA2ox1 and At GA2ox2, GA deactivation genes, is increased with addition of GA.

REGULATION

Regulation by other Hormones

The auxin indole-3-acetic acid (IAA) regulates concentration of GA1 in elongating internodes in peas. Removal of IAA by removal of the apical bud, the auxin source, reduces the concentration of GA1, and reintroduction of IAA reverses these effects to increase the concentration of GA1. This phenomenon has also been observed in tobacco plants. Auxin increases GA 3-oxidation and decreases GA 2-oxidation in barley. Auxin also regulates GA biosynthesis during fruit development in peas. These discoveries in different plant species suggest the auxin regulation of GA metabolism may be a universal mechanism.

Ethylene Decreases the Concentration of Bioactive GAs.

Regulation by Environmental Factors

Recent evidence suggests fluctuations in GA concentration influence light-regulated seed germination, photomorphogenesis during de-etiolation, and photoperiod regulation of stem elongation and flowering. Microarray analysis showed about one fourth cold-responsive genes are related to GA-regulated genes, which suggests GA influences response to cold temperatures. Plants reduce growth rate when exposed to stress. A relationship between GA levels and amount of stress experienced has been suggested in barley.

Role in seed Development

Bioactive GAs and abcisic acid levels have an inverse relationship and regulate seed development and germination. Levels of FUS3, an Arabidopsis transcription factor, are upregulated by ABA and downregulated by GA, which suggests that there is a regulation loop that establishes the balance of GA and ABA.

SIGNALLING MECHANISM

Receptor

In the early 1990's, there were several lines of evidence that suggested the existence of a GA receptor in oat seeds that was located at the plasma membrane. However despite intensive research, to date, no membrane-bound GA receptor has been isolated. This, along with the discovery of a soluble receptor, GA INSENSITIVE DWARF 1 (GID1) has led many to doubt that a membrane-bound receptor exists.

GID1 was first identified in rice and in Arabidopsis there are three orthologs of GID1, AtGID1a, b, and c. GID1s have a high affinity for bioactive GAs. GA binds to a specific binding pocket on GID1; the C3-hydoxyl on GA makes contact with tyrosine-31 in the GID1 binding pocket. GA binding to GID1 causes changes in GID1 structure, causing a 'lid' on GID1 to cover the GA binding pocket. The movement of this lid results in the exposure of a surface which enables the binding of GID1 to DELLA proteins.

DELLA Proteins: Repression of a Repressor

DELLA proteins, such as SLR1 in rice or GAI and RGA in Arabidopsis are repressors of plant development. DELLAs inhibit seed germination, seed growth, flowering and GA reverses these effects. DELLA proteins are characterized by the presence of a DELLA motif (aspartate-glutamate-leucine-leucine-alanine or D-E-L-L-A in the single letter amino acid code).

When GA binds to the GID1 receptor, it enhances the interaction between GID1 and DELLA proteins, forming a GA-GID1-DELLA complex. When in the GA-GID1-DELLA complex, it is thought that DELLA proteins undergo changes in structure that enable their binding to F-box proteins (SLY1 in arabidopsis or GID2 in rice). F-box proteins catalyse the addition of ubiquitin to their targets. The addition of ubiquitin to DELLA proteins promotes their degradation via the 26S-proteosome. The degradation of DELLA proteins releases cells from their repressive effects.

Targets of DELLA Proteins

Transcription Factors

The first targets of DELLA proteins identified were PHYTOCHROME INTERACTING FACTORs (PIFs). PIFs are transcription factors that negatively regulate light signalling and are strong promoters of elongation growth. In the presence of GA, DELLAs are degraded and this then allows PIFs to promote elongation. It was later found that DELLAs repress a large number of other transcription factors, among which are positive regulators of auxin, brassinosteriod and ethylene signalling. DELLAs can repress transcription factors either by stopping their binding to DNA or by promoting their degradation.

Prefoldins and microtubule assembly

In addition to repressing transcription factors, DELLAs also bind to prefoldins (PFDs). PFDs are molecular chaperones, meaning they assist in the folding of other proteins. PFDs function in the cytosol but when DELLAs bind to PFDs, it restricts them to the nucleus. An important function of PFDs is to assist in the folding of â-tubulin. As such, in the absence of GA (when there is a high level of DELLA proteins), PDF function is reduced and there is a lower cellular pool of â-tubulin. When GA is present the DELLAs are degraded, PDFs can move to the cytosol and assist in the folding of â-tubulin. â-tubulin is a vital component of the cytoskeleton (in the form of microtubules). As such, GA allows for re-organisation of the cytoskeleton, and the elongation of cells. Microtubules are also required for the trafficking of membrane vesicles. Membrane vesicle trafficking is needed for the correct positioning of several hormone transporters. One of the most well characterized hormone transporters arePIN proteins, which are responsible for the movement of the hormone auxin between cells. In the absence of GA, DELLA proteins reduce the levels of microtubules and thereby inhibit membrane vesicle trafficking. This reduces the level of PIN proteins at the cell membrane, and the level of auxin in the cell. GA reverses this process and allows for PIN protein trafficking to the cell membrane to enhance the level of auxin in the cell.

Discovery

It is the component responsible for the 'bakane' disease of rice seedlings. The disease is caused by the fungal pathogen Gibberella fujikuroi. E. Kurosawa treated uninfected rice seedlings with sterile filtrates of the fungus and reported the appearance of disease symptoms. Finally, the active substance causing the disease was identified as gibberellic acid.

Types

There exist more than 100 gibberellins obtained from a variety of organisms from fungi to higher plants. They are all acidic and are denoted as follows – GA1, GA2, GA3 etc. GA3 (Gibberellic acid) is the most noteworthy since it was the first to be discovered and is the most studied.

Effects

- Increase the axis length in plants such as grape stalks.
- Delay senescence (i.e. ageing) in fruits. As a result, their market period is extended.
- Help fruits like apples to elongate and improve their shape.

Applications

- The brewing industry uses GA3 to speed the malting process.
- Spraying gibberellins increase sugarcane yield by lengthening the stem.
- Used to hasten the maturity period in young conifers and promote early seed production.
- to promote bolting (i.e. sudden growth of a plant just before flowering) in cabbages and beet.

Cytokinins

Cytokinins (CK) are a class of plant growth substances (phytohormones) that promote cell division, or cytokinesis, in plant roots and shoots. They are involved primarily in cell growth and differentiation, but also affect apical dominance, axillary bud growth, and leaf senescence. Folke Skoog discovered their effects using coconut milk in the 1940s at the University of Wisconsin–Madison.

There are two types of cytokinins: adenine-type cytokinins represented by kinetin, zeatin, and 6-benzylaminopurine, and phenylurea-type cytokinins like diphenylurea and thidiazuron (TDZ). Most adenine-type cytokinins are synthesized in roots. Cambium and other actively dividing tissues also synthesize cytokinins. No phenylurea cytokinins have been found in plants. Cytokinins participate in local and long-distance signalling, with the same transport mechanism as purines and nucleosides. Typically, cytokinins are transported in the xylem.

Cytokinins act in concert with auxin, another plant growth hormone. The two are complementary, having generally opposite effects. The ratio of auxin to cytokinin plays an important role in the effect of cytokinin on plant growth. Cytokinin alone has no effect on parenchyma cells. When cultured with auxin but no cytokinin, they grow large but do not divide. When cytokinin is added, the cells expand and differentiate. When cytokinin and auxin are present in equal levels, the parenchyma cells form an undifferentiated callus. More cytokinin induces growth of shoot buds, while more auxin induces root formation.

Cytokinins are involved in many plant processes, including cell division and shoot and root morphogenesis. They are known to regulate axillary bud growth and apical dominance. The "direct inhibition hypothesis" posits that these effects result from the cytokinin to auxin ratio. This theory states that auxin from apical buds travels down shoots to inhibit axiliary bud growth. This promotes shoot growth, and restricts lateral branching. Cytokinin moves from the roots into the shoots, eventually signaling lateral bud growth. Simple experiments support this theory. When the apical bud is removed, the axillary buds are uninhibited, lateral growth increases, and plants become bushier. Applying auxin to the cut stem again inhibits lateral dominance.

While cytokinin action in vascular plants is described as pleiotropic, this class of plant hormones specifically induces the transition from apical growth to growth via a three-faced apical cell in mossprotonema. This bud induction can be pinpointed to differentiation of a specific single cell, and thus is a very specific effect of cytokinin.

Cytokinins have been shown to slow aging of plant organs by preventing protein breakdown, activating protein synthesis, and assembling nutrients from nearby tissues. A study that regulated leaf senescence in tobacco leaves found that wild-type leaves yellowed while transgenic leaves remained mostly green. It was hypothesized that cytokinin may affect enzymes that regulate protein synthesis and degradation. Cytokinin signaling in plants is mediated by a two-component phosphorelay. This pathway is initiated by cytokinin binding to a histidine kinase receptor in the endoplasmic reticulum membrane. This results in the autophosphorylation of the receptor, with the phosphate then being transferred to a phosphotransfer protein. The phosphotransfer proteins can then phosphorylate the type-B response regulators (RR) which are a family of transcriptions factors. The phosphorylated, and thus activated, type-B RRs regulate the transcription of numerous genes, including the type-A RRs. The type-A RRs negatively regulate the pathway.

Biosynthesis

Adenosine phosphate-isopentenyltransferase (IPT) catalyses the first reaction in the biosynthesis of isoprene cytokinins. It may use ATP, ADP, or AMP as substrates and may use dimethylallyl pyrophosphate (DMAPP) or hydroxymethylbutenyl pyrophosphate (HMBPP) as prenyl donors. This reaction is the rate-limiting step in cytokinin biosynthesis. DMADP and HMBDP used in cytokinin biosynthesis are produced by the methylerythritol phosphate pathway (MEP). Cytokinins can also be produced by recycled tRNAs in plants and bacteria. tRNAs with anticodons that start with a uridine and carrying an already-prenylated adenosine adjacent to the anticodon release on degradation the adenosine as a cytokinin. The prenylation of these adenines is carried out by tRNA-isopentenyltransferase.

Auxin is Known to Regulate the Biosynthesis of Cytokinin.

Uses

Because cytokinin promotes plant cell division and growth, produce farmers use it to increase crops. One study found that applying cytokinin to cotton seedlings led to a 5–10% yield increase under drought conditions. Cytokinins have recently been found to play a role in plant pathogenesis. For example, cytokinins have been described to induce resistance against Pseudomonas syringae in Arabidopsis thaliana and Nicotiana tabacum. Also in context of biological control of plant diseases cytokinins seem to have potential functions. Production of cytokinins by Pseudomonas fluorescens G20-18 has been identified as a key determinant to efficiently control the infection of A. thaliana with P. syringae..

Discovery

F. Skoog and his co-workers observed a mass of cells called 'callus' in tobacco plants. These cells proliferated only when the nutrient medium contained auxins along with yeast extract or extracts of vascular tissue. Skoog and Miller later identified the active substance responsible for proliferation and called it kinetin.

Types

Cytokinins were discovered as kinetin. Kinetin does not occur naturally but scientists later discovered several natural (example – zeatin) and synthetic cytokinins. Natural cytokinins exist in root apices and developing shoot buds – areas where rapid cell division takes place.

Effects

- Help in the formation of new leaves and chloroplast.
- Promote lateral shoot growth and adventitious shoot formation.
- Help overcome apical dominance.
- Promote nutrient mobilisation which in turn helps delay leaf senescence.

ABSCISIC ACID

In plants

Function

ABA was originally believed to be involved in abscission. This is now known to be the case only in a small number of plants. ABA-mediated signaling also plays an important part in plant responses to environmental stress and plant pathogens. The plant genes for ABA biosynthesis and sequence of the pathway have been elucidated. ABA is also produced by some plant pathogenic fungi via a biosynthetic route different from ABA biosynthesis in plants.

Abscisic acid owes its names to its role in the abscission of plant leaves. In preparation for winter, ABA is produced in terminal buds. This slows plant growth and directs leaf primordia to develop scales to protect the dormant buds during the cold season. ABA also inhibits the division of cells in the vascular cambium, adjusting to cold conditions in the winter by suspending primary and secondary growth.

Abscisic acid is also produced in the roots in response to decreased soil water potential (which is associated with dry soil) and other situations in which the plant may be under stress. ABA then translocates to the leaves, where it rapidly alters the osmotic potential of stomatal guard cells, causing them to shrink and stomata to close. The ABA-indced stomatal closure reduces transpiration (evaporation of water out of the stomata), thus preventing further water loss from the leaves in times of low water availability. A close linear correlation was found between the ABA content of the leaves and their conductance (stomatal resistance) on a leaf area basis.

Seed germination is inhibited by ABA in antagonism with gibberellin. ABA also prevents loss of seed dormancy. Several ABA-mutantArabidopsis thaliana plants have been identified and are available from the Nottingham Arabidopsis Stock Centre - both those deficient in ABA production and those with altered sensitivity to its action. Plants that are hypersensitive or insensitive to ABA show phenotypes in seed dormancy, germination, stomatal regulation, and some mutants show stunted growth and brown/yellow leaves. These mutants reflect the importance of ABA in seed germination and early embryo development.

Pyrabactin is a naphthalene sulfonamidehypocotyl cell expansion inhibitor, which is an agonist of the seed ABA signaling pathway. It is the first agonist of the ABA pathway that is not structurally related to ABA.

HOMEOSTASIS

Biosyn Thesis

Abscisic acid (ABA) is an isoprenoid plant hormone, which is synthesized in the plastidal2-C-methyl-D-erythritol-4-phosphate (MEP) pathway; unlike the structurally related sesquiterpenes, which are formed from the mevalonic acid-derived precursor farnesyl diphosphate (FDP), the C15 backbone of ABA is formed after cleavage of C40carotenoids in MEP. Zeaxanthin is the first committed ABA precursor; a series of enzyme-catalyzed epoxidations and isomerizations via violaxanthin, and final cleavage of the C40carotenoid by a dioxygenation reaction yields the proximal ABA precursor, xanthoxin, which is then further oxidized to ABA. viaabscisic aldehyde.

Abamine has been designed, synthesized, developed and then patented as the first specific ABA biosynthesis inhibitor, which makes it possible to regulate endogenous levels of ABA.

Location and timing of ABA biosynthesis

- Released during desiccation of the vegetative tissues and when roots encounter soil compaction.
- Synthesized in green fruits at the beginning of the winter period

- Synthesized in maturing seeds, establishing dormancy
- Mobile within the leaf and can be rapidly translocated from the roots to the leaves by the transpiration stream in the xylem
- Produced in response to environmental stress, such as heat stress, water stress, salt stress
- Synthesized in all plant parts, e.g., roots, flowers, leaves and stems
- ABA is synthesised in almost all cells that contain chloroplasts or amyloplasts

Inactivation

ABA can be catabolized to phaseic acid via CYP707A (a group of P450 enzymes) or inactivated by glucose conjugation (ABA-glucose ester) via the enzyme AOG. Catabolism via the CYP707As is very important for ABA homeostasis, and mutants in those genes generally accumulate higher levels of ABA than lines overexpressing ABA biosynthetic genes. In soil bacteria, an alternative catabolic pathway leading to dehydrovomifoliol via the enzyme vomifoliol dehydrogenase has been reported.

Effects

- Antitranspirant - Induces stomatal closure, decreasing transpiration to prevent water loss.
- Inhibits fruit ripening
- Responsible for seed dormancy by inhibiting cell growth – inhibits seed germination
- Inhibits the synthesis of Kinetin nucleotide
- Downregulates enzymes needed for photosynthesis.
- Acts on endodermis to prevent growth of roots when exposed to salty conditions
- Delays cell division

The absence of ABA, the phosphatase ABI1-INSENSITIVE1 (ABI1) inhibits the action of SNF1-related protein kinases (subfamily 2) (SnRK2s). ABA is perceived by the PYRABACTIN RESISTANCE 1 (PYR1) and PYR1-like membrane proteins. On ABA binding, PYR1 binds to and inhibits ABI1. When SnRK2s are released from inhibition, they activate several transcription factors from the ABA RESPONSIVE ELEMENT-BINDING FACTOR (ABF) family. ABFs then go on to cause changes in the expression of a large number of genes. Around 10% of plant genes are thought to be regulated by ABA.

In fungi

Like plants, some fungal species (for example Cercospora rosicola, Botrytis cinerea and Magnaporthe oryzae) have an endogenous biosynthesis pathway for ABA. In fungi, it seems to be the MVA biosynthetic pathway that is predominant (rather than the MEP pathway that is responsible for ABA biosynthesis in plants). One role of ABA produced by these pathogens seems to be to suppress the plant immune responses.

Measurement of ABA concentration

Several methods can help to quantify the concentration of abscisic acid in a variety of plant tissue. The quantitative methods used are based on HPLC and GC, and ELISA. Recently, 2 independent FRET probes have been developed that can measure intracellular ABA concentrations in real time in vivo.

Discovery

Three independent researchers reported the purification and characterization of three different inhibitors – Inhibitor B, Abscission II and Dormin. Later, it was found that all three inhibitors were chemically identical and were, therefore, together were given the name abscisic acid. Abscisic acid mostly acts as an antagonist to Gibberellic acid.

Effects

- Regulate abscission and dormancy.
- Inhibit plant growth, metabolism and seed germination.
- Stimulates closure of stomata in the epidermis.
- It increases the tolerance of plants to different kinds of stress and is, therefore, called 'stress hormone'.
- Important for seed development and maturation.
- induces dormancy in seeds and helps them withstand desiccation and other unfavourable growth factors.

ETHYLENE

Discovery

A group of cousins showed that a gaseous substance released from ripe oranges hastens the ripening of unripe oranges. Consequently, they found that the substance was ethylene – a simple gaseous Plant Growth Regulator. Ripening fruits and tissues undergoing senescence produce ethylene in large amounts.

Effects

- Affects horizontal growth of seedlings and swelling of the axis in dicot seedlings.
- Promotes abscission and senescence, especially of leaves and flowers.
- Enhances respiration rate during ripening of fruits. This phenomenon is 'respiratory climactic'.
- Increases root growth and root hair formation, therefore helping plants to increase their absorption surface area.

Application

Ethylene regulates many physiological processes and is, therefore, widely used in agriculture. The most commonly used source of ethylene is Ethephon. Plants can easily absorb and transport an aqueous solution of ethephon and release ethylene slowly.

- Used to break seed and bud dormancy and initiate germination in peanut seeds.
- To promote sprouting of potato tubers.
- Used to boost rapid petiole elongation in deep water rice plants.
- To initiate flowering and synchronising fruit-set in pineapples.
- To induce flowering in mango.
- Ethephon hastens fruit ripening in apples and tomatoes and increases yield by promoting female flowering in cucumbers. It also accelerates abscission in cherry, walnut and cotton.

3

BIOTIC AND ABIOTIC STRESS MANAGEMENT IN FRUIT CROPS

Selection of Suitable Crops and Varieties

In vegetable crops like dolichos bean, cowpea, cluster bean, lima bean, chilli, drumstick, brinjal, okra are suitable for rain-fed cultivation. Among these, legume vegetables can be recommended for contingency crop-planning in an eventuality of late monsoon rains. Varieties having good root system and capacity to recoup after the alleviation of stress need to be selected. Depending upon situation, it is recommended to use short duration varieties. Various factors should be considered in crop selection. This is a requisite that must be undertaken before actually starting a farming venture.

Even without a predetermined location and site of a farm, the crop to be grown can be decided though based mainly on its marketability and profitability. However, there are many cases especially in countries with agriculture-based economy in which the farm lot is already available. It could have been acquired through inheritance, or by purchase, or otherwise transferred through other means. Whatsoever, crop and varietal selection is the first consideration in starting or developing the farm. Right decision in the selection of crop or crops to be grown, particularly perennial types, will ultimately convert into a successful farming venture. Despite low profitability, many farmers with limited capital prefer to plant coconut due to various advantages such as readily available planting materials at low or no cost, ease in growing, available labor, low cost of production, compatibility with many intercrops, and a ready market

Save

In addition to the purpose of farming, the major factors to consider in crop selection include the following:

Prevailing farm conditions. An environmental scanning should first be conducted. This involves a thorough ocular inspection and other methods to obtain information on the biotics factor that can affect plant growth and yield, soil and climatic conditions prevailing in the area, and accessibility. Here the guiding rule is: recognize your farm initial then choose the correct crop

The biotic factors refer to living organisms including ruminant animals, insect and other pests, disease pathogens and weeds, as well as organisms having beneficial effects like civet cat population for the production of civet coffee and the abundance of pollinators. Where there is prevalence of a disease in a locality, susceptible crops may be excluded or a resistant variety may be selected.

The topographic features of the land like elevation, slope, terrain and aspect (the direction that a sloping land faces); and the physical and chemical properties of the soil such as texture, color, organic matter content, pH and fertility levels will determine the crops that are naturally suited. Also, the various climatic factors, such as prevailing climate type, temperature, rainfall, relative humidity, incidence of light, and frequency of typhoons will limit the choice of crops. A stable supply of water within the farm will allow wide possibilities in crop selection.

In addition, the accessibility of the farm to and from the market will influence the choice of crops. For example, cassava and oil palm should be preferably grown in farms with good roads and as close as possible to the market because the harvest is bulky and must be transported immediately due to rapid rate of degradation. Having a farm-to-market road is likewise important in stocking the farm with supplies. The crop(s) and the variety (ies) to be grown should be selected based on their adaptability to the prevailing conditions in the farm. A useful guide is to identify the crops growing in the farm and in the neighborhood. An interview of the neighboring farmers will also provide valuable information as to the probability of success, or failure, of growing certain preferred crops.

1. Marketability and profitability. For those who want to engage in cash crop farming or, at the least, ensure financial sustainability, crop selection must consider marketability and profitability. In general, this means that the crop to be selected must be high yielding. The product, be it the fruit,

seed, modified root or stem, flower or foliage or any part, must have an accessible, stable and robust market. With efficient labor and use of inputs, the harvest will realize profit to finance the suceeding farm activities or generate substantial return on investment. However, market and price are dictated by many factors such as the number of competitors, supply and demand, development of new products, promotional campaign, and agribusiness cycle.

2. Resistance to pests and diseases. Regardless of the purpose of farming, it is important to be able to select a crop and variety with wide resistance to important pests and diseases. The use of susceptible varieties may result to high cost of production or, worst, total crop failure.
3. Available technology. The technology for the growing of the crop must have been well established or easy to learn and apply. Likewise, certain crops are preferred because technical assistance is available locally.
4. Farming system. Crop selection is affected by the system of farming employed, that is, whether purely crop farming or integrated with livestock animals. Likewise, the particular crop species to be grown will depend on the crop production practices such as monoculture, multiple cropping, hedge row-strip cropping, and planting patterns.

   If there is a plan to integrate crop production with free-range livestock, or where entry of astray animals is unhampered, it would be wise to install tree guards or fences. However, additional cost can be eliminated or minimized by selecting crops that are less susceptible to nibbling. Some crops, like soursop (or guyabano) and sugar apple (or atis) have anti-herbivory properties. In intercropping, it is desirable that the component crops have complementary or mutualistic relation.
5. Security. In the absence of security personnel or where there is no fence that will exclude intruders, crop selection may be done in favor of those which are not susceptible of thievery. This will avoid such crops as vegetables and fruit crops which can be easily harvested for food and for cash.
6. Availability and cost of planting materials. These are important considerations especially among those who have limited capital. Take for example African oil palm. Even assuming that the projected return on investment is high and net profit can be realized in a short period of time, it would be difficult for small farmers to buy certified seedlings where the cost is about 10 times that of coconut. Plus the cost of hauling may be staggering.

Another is macapuno coconut. There are only a few tissue culture laboratories which propagate seedlings capable of producing up to 100% macapuno nuts per bunch and these laboratories may be too far away. The cost of producing seedlings being highly technical and costly, it is likewise expected that buying price would be too high. The last time that the herein author inquired, price per seedling was about 20 times more than that of normal coconut. This is one main reason why crop selection favors the normal varieties of coconut.

8. Availability and cost of labor. Where farm laborers, particularly in the community where the farm is located, are scant or cost is high, proper crop selection becomes more important. The need for farm workers should be a prime consideration because there is continuous need from land preparation to planting and up to harvest, post harvest, and marketing. For perennial crops, in fact, the work does not end with the first harvest but rather perpetually.

In such case, it may be preferable to select crops which do not require plenty of farm workers at regular basis. Or, at least, avoid those crops which are heavily dependent on regular labor, for example cacao over coffee or coconut.

Otherwise the farm becomes untended and the fruits rot or otherwise become waste. Likely an error in crop selection

Improved Method of Seedling Production

Improved method of seedling production such as Protray grown seedling using coco peat, nylon net protection and bio-fertilizers/bio-pesticide inoculation at nursery stage has good potential for obtaining sturdy, uniform and healthy seedlings. These seedlings when transplanted in the main-field will establish better with less root damage and fare better in overcoming biotic and abiotic stresses particularly during water stress conditions. In traditional Asian agriculture, farmers saved the seed from their own crops to use the following year. Now that most farmers are producing for commercial production, they are buying seed of improved varieties. A good supply of reliable, high-quality seed is a basic need for the development of Asian farming. New technical developments are helping produce better seed at a lower cost. A good harvest is impossible without good seed. For agriculture to prosper, farmers must have a reliable supply of high-quality seeds and seedlings, at an affordable price. Fortunately, recent advances in the technology of seed and seedling production are helping to improve both the quality and range of planting materials in Asia. These were discussed in a Workshop held by FFTC in June 2001. Some of the main points are summarized below.

BIOTECHNOLOGY

Tissue Culture

These were discussed in a Workshop held by FFTC in June 2001. Some of the main points are summarized below. Tissue culture of vegetables and other crops has been carried out for twenty years. Scientists are still finding new ways in which this technology can be applied to Asian agriculture. One of the most important is in the production of disease-free seedlings. Plants infected by virus always suffer some loss of vigor. However, there may not be any visible symptoms in seedlings. Several Asian countries are using tissue culture to mass produce virus-free banana seedlings.

Tissue culture is also being used in Taiwan to produce large-numbers of disease-free medicinal plants. Many of these belong to species which are becoming rare in the wild. This program serves the double purpose of conserving rare plants, as well as providing farmers with new, high-value crops.

Dna Technology

DNA techniques using primers offer a new method of detecting viruses and other plant diseases. Although this method is very accurate, it is also costly. It is used mainly for high-value crops which are propagated vegetatively, such as orchids.

DNA fingerprinting can be used to characterize the genetic make-up of the crop. In the past, this could only be done by sowing the seed and describing the physical characteristics of the growing plants. This took time, and was not very accurate. DNA methods tell us the genetic characteristics of the plant much more accurately, using DNA extracted from the seed or young seedling. This is useful in a number of different ways. It helps plant breeders produce varieties which have the specific qualities which farmers and consumers are looking for. It also protects the intellectual property rights of plant breeders. Another useful function is in the maintenance of germplasm banks. Purity and trueness of variety are very important for germplasm collections. DNA fingerprinting gives an accurate and detailed analysis of the genetic make-up of each accession. It can also be used to show whether genetic change is taking place over time. Seeds in germplasm collections must be repeatedly sown and harvested, so that the seed stays viable. In doing this, there is a danger that genetic drift will occur. While it is impossible to halt genetic drift completely, DNA methods allow us to detect whether it is taking place, and if so, ‘ to what extent.

THE DNA-TEC LABORATORY, THAILAND

A major problem in using DNA technology is that the laboratory facilities needed are very expensive. At the FFTC Workshop, the participants showed great interest in Thailand's "DNA-TEC".

This is a service laboratory established by the Thai Government. It serves both public research stations and private commercial companies. For the first few years it had government support, but it is now financially self-supporting.

The DNA-TEC laboratory gives small seed companies and government plant breeders access to DNA technology, at a price they can afford. An important recommendation of the FFTC workshop was that this should be explored as a model for other countries in the region. FFTC and the sponsoring organizations were also asked to consider organizing training courses in molecular marker technologies.

Hybrid Seed

Hybrid seed makes use of the well-known trait of "hybrid vigor". This is a very important genetic trait. Hybrids of rice and other crops can be expected to give a yield 20-25% higher than ordinary varieties. In China, about 15 million hectares of paddy fields are now planted in hybrid rice. So are a growing number of rice paddies in Vietnam, India and the Philippines.

In the production of hybrid rice, a male sterile line is used. In conventional hybrid rice breeding, this generally involves a three-line system, using a sterile male line and two other lines. Recently a two-line system has been adopted in China. This uses a seed parent which is sterile in some environments, and fertile in others. A major drawback of hybrid seed is that farmers cannot save seed from their own crop to plant the following year. Instead, they must purchase seed every year. Current research is developing vegetative propagation of hybrid rice for resource poor farmers, such as ratoon crops. Hybrid seed is also labor intensive. To produce hybrid rice seed takes an extra 50 man-days per hectare. This is a disadvantage only in countries with a labor shortage. It could be seen as a benefit in countries where jobs are scarce in rural areas.

Transgenic Crops

Transgenic crops are still controversial, although the area planted in transgenic crops is increasing each year. In 1999, 43 million hectares of transgenic crops were grown in different countries of the world. More than three-quarters of these were grown in industrialized countries, particularly the United States. However, an increasing percentage are being planted in developing countries.

Whatever the advantages of transgenic crops, the technology of producing them will only be economically viable if there is market demand for the product. At the moment, consumer resistance to transgenic food crops is strong, especially in Europe. This in turn has limited the use of transgenic crops in Asia and the Pacific, particularly in food exporting nations such as Thailand. Whether consumer resistance will increase in time, or whether disappear, is difficult to predict. Most Asian countries are adopting a "wait-and-see" attitude, before deciding to invest a lot of money in transgenic crops. Since transgenic crops are new, the long-term effect of consuming them is not yet known. In industrialized countries with a food surplus, consumers might reasonably ask why they need to take such a risk. They are likely to prefer crops bred by conventional means. It is a different situation in countries where there is a food shortage. Here, the greatest risk for many people is hunger. Transgenic crops may play an important role in relieving future food shortages, especially if the emphasis is on producing transgenic crops resistant to stresses such as saline soils, or common plant diseases.

Conclusion

Recent advances in seed production are making it possible to characterize seeds in detail, so varieties can be tailored to meet the specific needs of different environments and markets. Biotechnology offers improved methods of testing seed for quality, including purity, distinctiveness and uniformity, without having to plant the crop. It gives improved methods of identifying and eliminating pathogens. It has also helped bring improved agronomic performance, crop quality and yields.

However, Dr. Noel Mamicpic, Director of APSA, emphasized that in using biotechnology, researchers and entrepreneurs must consider smallholder farmers, and how they may share in the benefits of these new techniques.

"Don't forget", said Dr. Mamicpic, "that the ultimate beneficiary must be the farmer, who will produce all our food."

Adoption of soil and Moisture Conservation Techniques

Contour cultivation, contour trip cropping, mixed Cropping, tillage, mulching, zero tillage, are some of the agronomical measures for the in-situ soil moisture conservation. Mechanical measures like contour bunding, graded bunding, bench terracing, vertical mulching etc. also need to be followed for effective soil and moisture conservation in dry lands. Another technology for efficient utilization of runoff is water harvesting recycling. Rainwater harvesting includes collecting runoff water into dug out ponds or tanks in small depressions, gullies and into storage dams of earth or masonry structures. Rain water harvesting is possible in areas having rainfall as little as 500 to 800 mm. Depending on the rainfall and soil characteristics, 10-50 % of the runoff can be collected in farm pond. Surface run off thus collected in a farm pond can be used to provide protective irrigation in the period of prolonged dry spell or through microirrigation techniques. In most developing countries, agriculture remains one of the largest sectors in the economy both in terms of its contributions to the GDP and generating employment (Shiferaw and Holden, 1999). Efficient soil management is a must, not only for agriculture, but also in development as it can have a huge impact on addressing issues of poverty and food security. Therefore it is substantial to exercise environmental rehabilitation, such as reducing soil erosion, for survival and development by implementing appropriate approaches and technology. Soil erosion hampers agricultural productivity by deteriorating soil quality. Even though the effects of soil erosion vary by management and location, soil erosion deteriorates the chemical property of soil by loss of organic matter and loss of minerals containing plant nutrients (Zougmore et al., 2009). Soil erosion also exposes subsoil minerals with low fertility or high acidity. Soil erosion causes changes in physical properties of soil such as texture, infiltration rate, bulk density, available water holding capacity, depth of favorable root growth, etc.

Intolerable soil erosion and its impact occurs when the practiced farming system fails to take into account the ease with which soil can be washed away. Due to inappropriate farming systems and less attention given to conservation, soil erosion has been one of the major problems in Africa. In the tropics, there were isolated accounts of erosion from local agricultural staff during the 1930s and 1940s and in 1945 soil conservation became part of the agricultural policy of the colonial period (Young, 1997). Since, decades, attempts have been made to tackle soil erosion problems using different approaches and programmes in many countries for sustainability of agricultural production. Watershed management, particularly Soil and Water Conservation (SWC), supports sustainable livelihoods through reducing environmental degradation and increasing crop production (as it increases infiltration and reduces erosion as well as maintains soil fertility).

For sustainability of livelihood, in some case profitability in any intervention is indispensable. To maintain farm incomes and reduce externalities associated with erosive agricultural techniques, considerable efforts have been directed toward identifying and promoting profitable soil conservation strategies in low-income countries (Shively, 1997). The use of conservation practices in agricultural production is determined by the different income-earning strategies themselves and in combination with some of the biophysical and socio-economic conditions

(Jansen et al., 2006). In the livelihood strategies improved production and productivity is the major target. Productivity and conservation objectives are highly complementary, because conservation of soil, water and natural vegetation leads to higher productivity of crops and livestock and thus improvement of livelihoods (Kerr, 2002). According to Kerr (2002), the complementary relationship between productivity and conservation explains the importance and popularity of watershed development among planners, but the relationship with poverty alleviation is ambiguous because net benefits are skewed toward the wealthiest landowners while the poorest people bear many of the costs.

Diminishing productivity due to soil erosion-induced degradation of agricultural land has been of great concern (Amsalu and de Graaff, 2006). Measurements from experimental plots and micro-watersheds in Ethiopia estimated an annual soil loss of about 42 t ha from croplands (Bekele and Drake, 2003). Soil erosion and the consequential land degradation, has been recognized to as a serious problem in Ethiopia since 1973-74, subsequent to the devastating famine of the time (Dejene, 2003; Bewket, 2007). In Ethiopia, the traditional stone bunds on agriculture land is commonly practiced since many century in the Tigray highlands and North Shewa.

In Ethiopia, significant SWC activities were implemented during the 1970 and 1980s by mobilizing farmers through their peasant associations, mainly in food for work programs (Bewket, 2007). This approach was criticized for its top down approach. In many parts of the country, the current government has also been undertaking SWC through integrated and participatory watershed development approaches to improve rural livelihoods with sustainable natural resource management. In the government plan of 2006-2011, Plan for Accelerated and Sustainable Development to End Poverty (PASDEP), one of the goals was to enhance food security through improved natural resources management (MoFED, 2006). The ongoing 30 day national SWC-based watershed management campaign which started in 2010/11 and is expected to continue, also indicates the motive of the government and farmers to conserve soil. This campaign mainly promotes and implements physical SWC measures such as level soil bund, level fanya juu, stone bund, etc and undertakes planting billions of seedlings every year at the national level.

Enhancing soil Organic Matter Content

Constant efforts must be made to improve the soil organic carbon. Incorporation of crop residues and farm yard manure to soil improves the organic matter status, improves soil structure and soil moisture storage capacity. Organic matter content of the soil can also be improved by fallowing alley cropping, green manuring, crop rotation and agro forestry. Vegetable being short duration crop and having faster growth phases, the available organic matter needs to be properly composted. Vermicomposting can be followed for quicker usage of available organic matter in the soil and improving the soil moisture holding capacity.

Application of Foliar Nutrition:

The foliar application of nutrients during water stress conditions helps in the better growth by quick absorption of nutrients. The spraying of K and Ca induces drought tolerance in vegetable crops. Spraying of micronutrients and secondary nutrients improves crop yields and quality.

Use of drip irrigation

Drip irrigation has proved its superiority over other conventional method of irrigation, in Horticulture due to precise and direct application of water in root zone. A considerable saving in water, increased growth, development and yields of fruits and vegetables and control of weeds, saving in labour under drip irrigation are the added advantages. Drip irrigation can be adopted in fruit crops and also to all vegetable crops including closed spaced crops like onions and beans.The saving in water is to the tune of 30-50 % depending on the crop and season. Generally inline drip laterals having emitting point spaced at 30cm distance and emitting at the rate of 2LPH is selected for vegetable crops. In crops like chilli, brinjal, cauliflower and okra paired row planting is practiced and one drip lateral is used for two crop rows. Primitive drip irrigation has been used since ancient times. Fan Shengzhi shu, written in China during the first century BCE, describes the use of buried, unglazed clay pots filled with water, sometimes referred to as Ollas, as a means of irrigation. Modern drip irrigation began its development in Germany in 1860 when researchers began experimenting with subsurface irrigation using clay pipe to create combination irrigation and drainage systems. The research was later expanded in the 1920s to include the application of perforated pipe systems. The usage of plastic to hold and distribute water in drip irrigation was later developed in Australia by Hannis Thill.

Usage of a plastic emitter in drip irrigation was developed in Israel by Simcha Blass and his son Yeshayahu. Instead of releasing water through tiny holes easily blocked by tiny particles, water was released through larger and longer passageways by using velocity to slow water inside a plastic emitter. The first experimental system of this type was established in 1959 by Blass who partnered later (1964) with Kibbutz Hatzerim to create an irrigation company called Netafim. Together they developed and patented the first practical surface drip irrigation emitter.

After its first introduction in California in the late 1960s, only 5% of irrigated land utilized this system by 1988. By 2010, 40% of irrigated land in California utilized this system. Modern drip irrigation has arguably become the world's most valued innovation in agriculture since the invention in the 1930s of the impact sprinkler, which offered the first practical alternative to surface irrigation. Drip irrigation may also use devices called micro-spray heads, which spray water in a small area, instead of dripping emitters. These are generally used on tree and vine crops with wider root zones. Subsurface drip irrigation (SDI) uses permanently or temporarily buried dripperline or drip tape located at or below the plant roots. It is becoming popular for row crop irrigation, especially in areas where water supplies are limited, or recycled water is used for irrigation. Careful study of all the relevant factors like land topography, soil, water, crop and agro-climatic conditions are needed to determine the most suitable drip irrigation system and components to be used in a specific installation.

Components and Operation

Components used in drip irrigation (listed in order from water source) include:

- Pump or pressurized water source
- Water filter(s) or filtration systems: sand separator, Fertigation systems (Venturi injector) and chemigation equipment (optional)
- Backwash controller (Backflow prevention device)
- Pressure Control Valve (pressure regulator)
- Distribution lines (main larger diameter pipe, maybe secondary smaller, pipe fittings)
- Hand-operated, electronic, or hydraulic control valves and safety valves
- Smaller diameter polyethylene tube (often called "laterals")
- Poly fittings and accessories (to make connections)
- Emitting devices at plants (emitter or dripper, micro spray head, inline dripper or inline drip tube)

In drip irrigation systems, pump and valves may be manually or automatically operated by a controller.

Most large drip irrigation systems employ some type of filter to prevent clogging of the small emitter flow path by small waterborne particles. New technologies are now being offered that minimize clogging. Some residential systems are installed without additional filters since potable water is already filtered at the water treatment plant. Virtually all drip irrigation equipment manufacturers recommend that filters be employed and generally will not honor warranties unless this is done. Last line filters just before the final delivery pipe are strongly recommended in addition to any other filtration system due to fine particle settlement and accidental insertion of particles in the intermediate lines.

Drip and subsurface drip irrigation is used almost exclusively when using recycled municipal wastewater. Regulations typically do not permit spraying water through the air that has not been fully treated to potable water standards.

Because of the way the water is applied in a drip system, traditional surface applications of timed-release fertilizer are sometimes ineffective, so drip systems often mix liquid fertilizer with the irrigation water. This is called fertigation; fertigation and chemigation (application of pesticides and other chemicals to periodically clean out the system, such as chlorine or sulfuric acid) use chemical injectors such as diaphragm pumps, piston pumps, or aspirators. The chemicals may be added constantly whenever the system is irrigating or at intervals. Fertilizer savings of up to 95% are being reported from recent university field tests using drip fertigation and slow water delivery as compared to timed-release and irrigation by micro spray heads.

Properly designed, installed, and managed, drip irrigation may help achieve water conservation by reducing evaporation and deep drainage when compared to other types of irrigation such as flood or overhead sprinklers since water can be more precisely applied to the plant roots. In addition, drip can eliminate many diseases that are spread through water contact with the foliage. Finally, in regions where water supplies are severely limited, there may be no actual water savings, but rather simply an increase in production while using the same amount of water as before. In very arid regions or on sandy soils, the preferred method is to apply the irrigation water as slowly as possible.

Pulsed irrigation is sometimes used to decrease the amount of water delivered to the plant at any one time, thus reducing runoff or deep percolation. Pulsed systems are typically expensive and require extensive maintenance. Therefore, the latest efforts by emitter manufacturers are focused on developing new technologies that deliver irrigation water at ultra-low flow rates, i.e. less than 1.0 liter per hour. Slow and even delivery further improves water use efficiency without incurring the expense and complexity of pulsed delivery equipment. An emitting pipe is a type of drip irrigation tubing with emitters pre-installed at the factory with specific distance and flow per hour as per crop distance.

An emitter restricts water flow passage through it, thus creating head loss required (to the extent of atmospheric pressure) in order to emit water in the form of droplets. This head loss is achieved by friction/turbulence within the emitter.

Advantages and disadvantages

Pressure compensated integral dripper on soilless without growing channels

The Advantages of drip Irrigation are:

- Fertilizer and nutrient loss is minimized due to a localized application and reduced leaching.
- Water application efficiency is high if managed correctly.
- Field leveling is not necessary.
- Fields with irregular shapes are easily accommodated.
- Recycled non-potable water can be safely used.
- Moisture within the root zone can be maintained at field capacity.
- Soil type plays a less important role in the frequency of irrigation.
- Soil erosion is lessened.
- Weed growth is lessened.
- Water distribution is highly uniform, controlled by the output of each nozzle.
- Labour cost is less than other irrigation methods.
- Variation in supply can be regulated by regulating the valves and drippers.
- Fertigation can easily be included with minimal waste of fertilizers.
- Foliage remains dry, reducing the risk of disease.
- Usually operated at lower pressure than other types of pressurized irrigation, reducing energy costs.
- The Disadvantages of drip Irrigation are:
- Initial cost can be more than overhead systems.
- The sun can affect the tubes used for drip irrigation, shortening their lifespan. (See Polymer degradation);
- The risks of degrading plastic affecting the soil content and food crops. With many types of plastic, when the sun degrades the plastic, causing it to become brittle, the estrogenic chemicals (that is, chemicals replicating female hormones) which would cause the plastic to retain flexibility have been released into the surrounding environment.

- If the water is not properly filtered and the equipment not properly maintained, it can result in clogging or bioclogging.
- For subsurface drip the irrigator cannot see the water that is applied. This may lead to the farmer either applying too much water (low efficiency) or an insufficient amount of water, this is particularly common for those with less experience with drip irrigation.
- Drip irrigation might be unsatisfactory if herbicides or top dressed fertilizers need sprinkler irrigation for activation.
- Drip tape causes extra cleanup costs after harvest. Users need to plan for drip tape winding, disposal, recycling or reuse.
- Waste of water, time and harvest, if not installed properly. These systems require careful study of all the relevant factors like land topography, soil, water, crop and agro-climatic conditions, and suitability of drip irrigation system and its components.
- In lighter soils subsurface drip may be unable to wet the soil surface for germination. Requires careful consideration of the installation depth.
- Most drip systems are designed for high efficiency, meaning little or no leaching fraction. Without sufficient leaching, salts applied with the irrigation water may build up in the root zone, usually at the edge of the wetting pattern. On the other hand, drip irrigation avoids the high capillary potential of traditional surface-applied irrigation, which can draw salt deposits up from deposits below.
- The PVC pipes often suffer from rodent damage, requiring replacement of the entire tube and increasing expenses.
- Drip irrigation systems cannot be used for damage control by night frosts (like in the case of sprinkler irrigation systems)

Drip tape

Drip tape is a type of thin-walled dripperline used in drip irrigation. The first drip tape was known as "Dew Hose". Drip tape duct tape is made of polyethylene and is sold flat on reels. The wall thickness typically ranges from 4 to 25 mils (0.1–0.6 mm). Thicker walled tapes are commonly used for permanent subsurface drip irrigation and thinner walled tapes for temporary throw-away type systems in high-value crops. Water exits from tape through emitters or drippers. The typical emitter spacing ranges from 6 to 24 inches (150–600 mm). In some products, the emitters are manufactured simultaneously with the tape and are actually formed as part of the product itself. In others, the emitters are manufactured separately and installed at the time of production. Some product is not a tape, but a thin-walled dripperline, but in popular parlance, both types of products are called tapes. Typical tape diameters are 5/8", 7/8", and 1-3/8", with the larger diameters more commonly used on permanent installations with longer runs. Drip tape is a recyclable material and can be recycled into viable plastic resins for reuse in the plastics manufacturing industry.

Irrigation dripper

Drip irrigation is used in farms, commercial greenhouses, and residential gardeners. Drip irrigation is adopted extensively in areas of acute water scarcity and especially for crops and trees such as coconuts, containerized landscape trees, grapes, bananas, ber, eggplant, citrus, strawberries, sugarcane, cotton, maize, and tomatoes. Drip irrigation for garden available in drip kits are increasingly popular for the homeowner and consist of a timer, hose and emitter. Hoses that are 4 mm in diameter are used to irrigate flower pots.

Use of micro sprinkler irrigation

Depending upon situation and availability of water, this technology can be used for fruits and vegetable crops. The cost of initial establishment is lower compared to drip system. Further in summer the sprinkling of water helps in reducing the microclimate temperature and increasing the humidity, thereby improving the growth and yield of the crop. The water saved is to the tune of 20 to 30 per cent. The micro-sprinklers are intended to provide irrigation using very fine droplets. They have a rotating deflector, called rotor or dancer, who helps deliver a greater diameter coverage, lower precipitation rate than the diffusers, increased droplet size, and better distribution of water (especially in distribution uniformity). For each type of micro-sprinkler are several types of rotors (dancers).

The main difference with nebulization is that microsprinckler projects water on tiny water jets to the plant instead of providing it in a nebulized way, and in turn they have rotating elements that distribute water to the surface.

It can be installed directly on PE pipe surface (25 or 35 mm in diameter) or may be arranged on support rods and microtube. It has an assembly and easy operation, without using tools for installation. All components are interchangeable, allowing you to use the most appropriate design for every need.

Applications

The micro sprinklers are ideal for low volume irrigation in horticultural crops, fruit, flowers, greenhouses, nurseries, frost protection and watering gardens.

They also allow the application of pesticides in plant cover crops. Its use is widespread in greenhouses, especially in leaf vegetables (lettuce, spinach, cabbage).

INSTALLATION PARTS:

- Pump assembly: to provide the appropriate pressure and flow rate to the facility.
- Filtering: The varying degrees of thickness thereof filtration anger related water quality, and the size of the sprinkler nozzle. .
- Fertilized System
- Pipe Network
- Micro-sprinklers: The scope, flow rate and droplet size will determine the choice of either model and mode of installation.

MATERIALS USED

Installing buried with PVC or PE and accessories needed.

Depending on the type of installation we find the following materials:

- Install air pipes PE, with the support system, microtubes and connection systems microsprinklers. (inverted microsprinkler).
- Installation of PE pipes in the ground, support rods, microtubes and connection systems microsprinklers.

Moisture saving methods under limited water resource conditions:

The following methods may be adopted under limited water conditions to save water:

Water saving irrigation method

Under limited water situations, water-saving irrigation methods like alternate furrow irrigation or widely spaced furrow irrigation and drip irrigation systems can be adopted. Studies conducted on methods of irrigation in capsicum, tomato, okra and cauliflower indicated that adopting alternate-furrow irrigation and widely-spaced furrow irrigation saved 35 to 40 per cent of irrigation water without adversely affecting yield.

MULCHING PRACTICES IN VEGETABLE PRODUCTION

The technique of covering the soil with natural crop residues or plastic films for soil and water conservation is called mulching. Mulching can be practiced in fruits and vegetable crops using crop residues and other organic material available in the farm. Recently plastic mulches have come into use due to the inherent advantages of efficient moisture conservation, weed suppression and maintenance of soil structure. Wide variety of vegetables can be successfully grown using mulches. In addition to soil and water conservation, improved yield and quality, suppression of weed growth , mulches can improve the use efficiency of applied fertilizer nutrients and also use of reflective mulches are likely to minimize the incidence of virus diseases. For vegetable production generally polyethylene mulch film of 30micron thick and 1 to 1.2 m width is used. Generally raised bed with drip irrigation system is followed while laying the mulch film.

Wind breaks, Hedges and Intercropping

To overcome the adverse effect of high temperature and dry winds, tall growing trees need to be planted all along the boundary of the farm. Inter cropping of vegetable crops of the area can be practiced in orchards during summer months. Maize/ Sorghum can be grown all along the border of the plot to mitigate the effect of desiccating winds.

Use of Protected Cultivation of Vegetables

In peri-urban regions where climate does not favour year round production of crops in the open field, vegetable production can be taken up in protected environment. Protective structure is a facility to protect crop from biotic and abiotic constraints. Structures for protected cultivation include green houses, plastic/net houses and "tunnels". Commonly used protected structures are polyhouses and net or shade houses. Rain-shelter is a simple structure covered with polyethylene sheet which helps in producing the crops which are affected by excessive rainfall. The productivity of tomato, onions and melons are adversely affected in the event of high rainfall due to difficulty in managing the foliar diseases, lack of proper soil aeration and drainage and also depending on the nature of crop physical damage of the foliage and flower drop. Net house cultivation and shade net cultivation provide better microclimate especially during summer in minimizing the high temperature effect and improving the relative humidity. The productivity of tomato, French bean and capsicum can be improved during high temperature period by using net/ shade net on the top.

Control of leaf Miner and mite during high Temperature Stress

For management of leaf miner spray neem soap 4 grams / liter or triazophos at the rate 1.5 ml / l. To manage mites spray Abanectin 0.5 ml/l. Aphids may be observed in case of beans. Spray neem soap (1.0 %) or neem seed kernel extract (4.00 %). Do's and Don'ts for Vegetable crops

1. Establishing vegetation-free strips under the trees and between tree rows before growth begins.
2. In crop like Onion, drum seeder may be used for direct sowing.
3. Postpone transplanting of seedlings in the main field and also fertilizer application till the favorable soil moisture is prevailed.
4. Once the soil moisture condition becomes favorable transplanting of the seedlings may be taken up.
5. Resorting to foliar application of (water soluble) major nutrients.
6. Protection of young plants with partial shade.
7. In the inter spaces of crops weeding and inter-culture practices may be followed.
8. Thinning may be taken up to reduce the plant population.
9. Alternate furrow Irrigation may be taken up based on the availability of water.
10. Drip Irrigation may be followed. Pitcher irrigation wherever drip is unavailable for protective maintenance.
11. Plastic mulching and drip irrigation may be followed for better soil and moisture conservation and weed control.
12. Adopting the conjunctive use of surface and ground water as well as the use of non-conventional sources such as brackish water.
13. Waste water should not be utilized without pretreatment & safe reuse may be ensured.
14. Minimise use of those fertilizers which promote vegetative growth like nitrogen, use K and B as foliar spray to maintain plant turgor.
15. Reducing water losses during conveyance and distribution.
16. Use of super absorbent polymers such as Luquasorb for water absorption and slow release.

Do's and Don'ts for Fruit Crops

1. In situ soil moisture conservation by trenching, contour/ field bunding, Gully Plugging, Loose boulder check dams may be taken up.
2. Use locally available organic mulches to conserve moisture in situ in the basin. Adopt to drip/ trickle irrigation for judicious use of water. In situ grafting on drought resistant root stocks for better establishment of saplings. Practicing conservation horticulture such as inter crops and soil moisture conservation.
3. Avoid soil application of fertilizers till sufficient soil moisture is available.
4. Adopting to foliar nutrition of major nutrients under water stress conditions to enhance the nutrient uptake and use efficiency.
5. Provide protective irrigation through pitcher and protective shade to young plants to reduce the high evaporative demand.
6. In addition to drip irrigation and mulching for production of fruit crops under water limiting conditions, novel irrigation methods, like partial root zone drying (PRD), could be adapted in grapes, mango and citrus. The partial root zone drying method helps in development of a deeper root system.
7. For all fruit crops, basin mulching with locally available plant material and plastic mulch may be taken up.
8. Try to compost all the available plant waste materials and use it as organic manure to fruit and vegetable crops.
9. For all the horticultural crops, drip Irrigation may be followed. Pitcher irrigation wherever drip is unavailable, is suitable for protective maintenance.

Cell trays are used by commercial growers to produce seedlings for planting out. The seedlings are easily removed from the tray for transplanting and the growth check to transplants from cell trays is minimal when planted in the field compared to the use of other types of transplants. Seedlings from cell trays may be used in manual or automatic transplanters. Soil-less mixes are usually used for commercial seedling production. Commercial sterilised soil-less mixtures are available with added fertiliser and this simplifies the production process.

Advantages and Disadvantages of using Transplants

Use of transplants has a number of advantages compared to direct seeding, although some vegetables are much more suited to growing as transplants than others. For example, beans, sweet corn, carrots and peas are amongst those that are not usually grown as transplants.

Advantages

- Shorter growing season and more efficient use of land.
- Improved crop uniformity.
- More accurate prediction of harvest date.
- Facilitates the use of a wider range of herbicides.
- Extends the growing season.
- More efficient use of expensive hybrid seed.

Disadvantages

- Transplant shock which delays growth but is not as severe on cell raised seedlings compared to bare rooted seedlings.
- High cost of seedlings which adds significantly to production cost.
- Extra labour to establish crop.
- Changes crop growth pattern.

Nursery Design

The capital cost needed to begin growing seedlings as cell transplants makes it more expensive than in seedbeds. The use of cell-grown plants have many advantages over seed bed or seed tray plants; a fact demonstrated by their widespread use.

Some flower and vegetable growers have set up their own cell production units. Growers must give the same careful attention to planting and design of the seedling growing area, just as if it were a commercial nursery.

Commercial nurseries constantly assess their production process to ensure they are only marketing quality seedlings. They seek to produce uniform, healthy plants in quantities that will meet their customers' demands.

Hygiene

Good nursery hygiene is essential if disease free seedlings are to be produced for transplanting in the field. Planning is essential to maintain nursery hygiene and to ensure a high standard of plant health. A seedling nursery must be clearly separated from any other growing system or area and there must be strict control over entry to the nursery to limit the possibility of introducing pests and diseases.

In other words, a nursery must be treated as a quarantine area, with restrictions on entry and movement inside the boundary. This applies to people and to materials or equipment. The design of the nursery should include defined areas of operation. Briefly these areas can be defined as:

- stores - general goods, chemicals and growing media. etc. as delivered.
- preparation - media mixing area including a section for sterilisation of media and trays; sterile material must be protected.'
- seeding house - includes a special seed store, seed treatment section, seeding and germination rooms.'
- pricking out - this is an extension of the germination room.,
- growing on - defined areas for different types and ages of plants.'
- preparation - a direct but isolated extension of the growing areas.'
- sales/dispatch - the only area open to the public.,
- recycling holding area for reject, surplus material or returned trays for recycling and composting., Until sterilised recycled materials need to be kept separate and isolated from the nursery .
- waste disposal - isolated, secure, well separated from seedling houses and downwind prior to collection or treatment for disposal.

It is essential to design a flow-through system to assist with hygiene controls. For example, a steam sterilisation kiln should have an entry and an exit door to a storage-holding area. So there is a clear separation of functions.

The provision of concrete aprons, floors and pathways designed for rapid drainage is more than just a sign of a successful nursery; it is essential for a healthy business.

Structures

Glass or plastic greenhouses or frames give the best protection to seedlings. Supplementary heating may be needed to promote healthy growth of warm season plants. Inflation of the air gap between a twin-skinned plastic house will help to improve heat retention inside the house. On hot and sunny days shading may be required along with proper ventilation and air circulation. Ceiling fans will assist in further air movement and in less condensation. Wind breaks will help reduce damage from a strong wind within the nursery area and help reduce the entry of disease-laden dust. They should be sited to protect the houses from the prevailing winds

Shade houses are used to harden greenhouse raised plants to conditions outside. They offer some protection from the extremes of temperature and from winds.

An alternative method to harden plants is to place them in an open area surrounded by a wind break such as shade cloth.

Growing mixtures

Hygiene of growing mixes is vital and it is important to ensure that all growing mixtures have been sterilised, especially soil based media.

Soil-based media

Soil has traditionally been accepted as part of seedling growing mixture. To most it would seem that the use of soil keeps seedling growing simple and straightforward.

This is not necessarily true. Consider these points:

- Quality standards will vary within and between loads of delivered soil. This applies to standards such as fineness or coarseness of the soil; the depth at which it was mined; and the amount of undesirable material coming with the soil such as diseases, weeds, trash and clay.
- Because the soil can vary, so can the desired standards of a growing mixture such as density, where balance is important for uniform strong plants. These differences may not be easily seen, but could be a reason for indifferent plant growth.
- Stock-piled soil can support unwanted plant growth (weeds) which can increase the amount of disease inoculum in the soil or the spread through a nursery.
- Soils may contain unknown plant toxins, extremes in soil pH variations, nutrient content or unknown chemical residues.

Soil-less media

The alternatives to soil, either in part or in total are not without problems. The worth of these materials and how they can be used must be understood.

Some soil-less Materials

- Pine bark: Two key disadvantages with pine bark are the effect of its toxins on plant growth and in its hydrophobic (or difficult to wet) nature. There have also been problems at times with purity of sample, and in uniformity of particle size following milling. For seedling mixes bark should be milled to a maximum particle size of 5 mm.Most of the toxin content can be leached by ageing in open air stacks.' however the bark must be wet first. Newly stripped bark is harder to wet following drying than stockpiled, weathered bark.Pine bark can be composted with the addition of small amounts for slow-release fertilisers, to help in the process. This process can give some control over certain diseases which may contaminate the bark. In open stacks', the heat generated throughout the heap will not be even and therefore in the cool areas, the temperature will not rise to the thermal death point of many diseases.

- Hardwood sawdust: There are similar problems with sawdust to those of pine bark, i.e. some timbers do have quite high toxin levels. Treatment is the same. Sawdust composted in the recommended manner can be combined with pine bark and with peat moss. It is desirable to keep the percentage by volume of sawdust to around 40 percent or less.
- Soft wood sawdust: The most common softwood sawdust available is from *Pinus radiata*, which contains less toxins than some of the Australian hardwoods.
- Sand: The value of adding sand, an inert mineral, to any seedling mix is to obtain a balance between the other components and to give some bulk to the final product. A common belief is that sand helps with drainage' in fact the reverse can occur. For example, if added to peat moss it may fill in many of the narrow spaces between peat particles, thereby reducing loss of moisture. Also re-wetting of a dried-out mix is improved if there is a small amount of sand present. However overmixing of peat and sand in a rotary mixer can lead to pulverising of the peat, thus changing its structural qualities. On the other hand large amounts of sand added to a seedling mix is likely to produce rather dense mixtures. Likewise there needs to be a balance between the extremes of fine and coarse sands. Sand 2 mm or larger is classed as coarse material.
- Perlite: As an inert light material, perlite is used to reduce bulk density . It does not retain moisture or hold any plant nutrient. There is no real value of having perlite as part of the growing media.
- Vermiculite: Horticultural vermiculite has long been valued by the nurse industry. It has most value in short-term mixes, such as those needed for seedlings. As a material with a high base exchange capacity, it is able to reduce the loss of nutrients through leaching. It also contains small amounts of magnesium and potassium.

- Polystyrene pellets or beads: This material is used sometimes, if for no other reason than to add bulk- It has no nutrient value or water holding capacity. Polystyrene is difficult to handle outside when there is a wind. although the results are better if mixed dry with a little brown coal, before moisture is added.
- Brown coal: This material is known also as Lignum Peat or coal fines. The ratio of brown coal in a soil-less mix to other components needs to be low, about 20 percent by volume or even less. Higher volumes with adequate nutrition may cause plants to respond with lush growth and will be difficult to harden. Brown coal has a high water-holding capacity, with a high level of unavailable water. Its water loss is in fact high too, and it has been shown that plants transpire less water growing in brown coal, than is the case with peat moss.
- Peat moss: The use of peat moss for horticulture is increasing world wide. In Australia there are a number of peat deposits, but not many are suited for seedling mixtures. A common local peat used in Victoria comes from Tasmania, but because of its trash content (old plant material) its quality does not compare with many imported peats. The fine sedge-peats are generally not suited to seedling mixtures due to their dense nature. Most peats are relatively stable against rapid decay. They are valued for their high water-holding capacity; the sphagnum peats having a better balance between water and air, than most sedge peats. Even though peats have this capacity to hold water, their loss of moisture is high through transpiration and evaporation. Because of this it is claimed that plants growing under optimum conditions would grow faster in this material.

Soil-less mixtures

There is no such thing as one perfect mix for all types of plants. With so many mixes to choose from it can be confusing.

Commercial mixes: Most nurseries buy pre-mixed commercial mixes and bulk lots of a prepared mix can be obtained with or without fertiliser. The standard for ingredients should never change unless by agreement.

Bulk commercial mixes simplify the process and fertiliser can be included, eliminating the need for adding fertiliser, lime etc to obtain a balanced mix.

The most universal is a 50:50 peat/vermiculite mix. Some nurseries use it successfully as a standard mix for a range of plant species.

When changes are made from a standard mix, never change more than one ingredient for each test. For example, if you change the 50:50 peat/vermiculite to a 60:40, do not add another ingredient until this change has been tested. This also applies to added fertilisers. In this way change is controlled and results understood.

Plant nutrition for soil-less mixes: During the final mixing, lime, gypsum, fine grade superphosphate, a trace element mix and small amounts of nitrogen and potassium are added. A fine grade slow-release fertiliser may be used instead of adding N.P.K. separately. Only enough nutrients should be added to the mix to obtain steady growth for the first 7 to 14 days. Further growth is maintained through foliar feeding with a soluble complete fertiliser every 7 to 10 days, until the seedlings are near maturity. Do not use fertiliser at higher than recommended rates.

Water plants with care because over-watering can easily leach nutrients out of soil-less mixes. Careful leaching of nutrients can be used as a growth control aid. Slowing seedling growth by this method is preferred to trimming the leaves of plants or with-holding irrigation before transplanting. Media pH: Lime has been normally added to correct a pH that is strongly acid. In fact the greatest benefit from lime and gypsum is the calcium content. Calcium, like phosphorus and other nutrients', is easily leached from soil-less mixes.

Seed health and quality

Healthy plants are encouraged by a balanced nutrition program giving steady growth. Excessive wetness of the leaf may promote disease. Poor ventilation or air circulation will add further to the problem. A healthy seed is a must. In some circumstances it may be advisable to treat the seed for diseases by using a seed coating treatment if this has not already been done by the seed supplier.

When using a seed dust always follow the instructions on the label and remove surplus dust from the seed and from the seed container or germination may be affected. For some vegetables a hot water treatment should be used to control seed borne diseases, but care must be taken as germination percentage can be reduced.

Only use seed from reliable sources. Do not accept grower seed or seed that is not sealed in a seed company container. Infected seed can introduce disease into the nursery. The results can be expensive and the disease difficult to control. Disease can lead to reduced quantities for orders or unnecessary production to make up for losses and hence tie-up expensive nursery space.

Quality begins with healthy seed; in seed vigour, and in high germination levels. Sometimes there are problems with seed, therefore it is essential that all sowings and seedling orders delivered are identified in a trace back system that includes date supplied, seed lot number, the date treated and sown, and the name of the seed company. When direct seeding only seed of the highest germination count should be used. For example, if the stated minimum germination is 85 per cent, then a minimum of 15 per cent more trays and greenhouse space will be required to meet an order.

Seeding

Air seeders are almost the only method used today in the production of cell-grown plants, because of their cost effectiveness. Such units vary from small hand operated single tray units to high volume, automated, flow-through systems. The high volume systems include auto tray-fill and topping of the sown tray. Plug seeding: This is a method where seed is sown into small plugs for germination. At an early stage the plug is slipped into the normal cell-tray to complete growth to a mature seedling. The advantage of this system is the huge saving in greenhouse space.

Irrigation

Use only quality water low in salt. Disease organisms may survive in water, so the source of the water must be considered carefully. Water should not be recycled due to the risk of infecting plants with diseases. The type of irrigation to achieve the most even growth is a travelling boom irrigator. It can also be used as an efficient management tool. The initial cost for a boom irrigator can be high if only looked at as an irrigation system. A boom irrigator becomes a management tool when it waters, feeds, applies most chemicals and is used to slow down plant growth. The whole process can be programmed through a computer taking into account different groups of plants. Irrigate only to the point of drainage, unless it is desired to slow growth down by leaching of some nutrients. Do not allow soil-less mixes to dry out to a point where plants wilt. Plant growth is affected and because over-watering is needed to re-wet the media in each cell, nutrients are leached away.

Pests and diseases

Hygiene as outlined under Nursery design has a vital role in the control of pests and diseases. Use steam or chemical sterilisation of the growing media, structures, tools and trays. Effective ventilation and air movement is also a sound disease prevention method. Individual crop Agriculture Notes should be consulted to understand pests and diseases that could affect the growth of healthy seedlings.

A guide to control measures will also be found in these Agriculture Notes. Care must be taken with the use of pesticides within enclosed areas. Note also that in a greenhouse plants can be more sensitive to chemicals than in the open field. Follow the directions on the label. If in doubt consult the chemical company named on the label.

Marketing

There are two choices in marketing the plants:

1. Trays of plants are sent out as orders. or
2. Plants are Pulled and Bulk delivered in Containers.

Extreme care is required where used trays or containers are returned to the nursery due to the risk of disease. Returned trays and containers should be kept in a separate area. Do not place these items along side or above trays containing plants or clean trays and containers. All equipment must first be sterilised before re-entering the main nursery complex.

Pulled plants may be stored in the cool room for two to three days, but should be planted as soon as possible. A high rejection rate of plants before delivery indicates a production problem which must be addressed and cannot be tolerated.

4

ADVANCES IN BREEDING OF FRUIT CROPS

MODERN PLANT BREEDING

Modern plant breeding may use techniques of molecular biology to select, or in the case of genetic modification, to insert, desirable traits into plants. Application of biotechnology or molecular biology is also known as molecular breeding.

Marker assisted selection

Sometimes many different genes can influence a desirable trait in plant breeding. The use of tools such as molecular markers or DNA fingerprinting can map thousands of genes. This allows plant breeders to screen large populations of plants for those that possess the trait of interest. The screening is based on the presence or absence of a certain gene as determined by laboratory procedures, rather than on the visual identification of the expressed trait in the plant.

The purpose of marker assisted selection, or plant genomes analysis, is to identify the location and function (phenotype) of various genes within the genome. If all of the genes are identified it leads to Genome sequence. All plants have varying sizes and lengths of genomes with genes that code for different proteins, but many are also the same. If a gene's location and function is identified in one plant species, a very similar gene likely can also be found in a similar location in another species genome.

Marker types

The majority of MAS work in the present era uses DNA-based markers. However, the first markers that allowed indirect selection of a trait of interest were morphological markers. In 1923, Sax first reported association of a simply inherited genetic marker with a quantitative trait in plants when he observed segregation of seed size associated with segregation for a seed coat color marker in beans (Phaseolus vulgaris L.). In 1935, Rasmusson demonstrated linkage of flowering time (a quantitative trait) in peas with a simply inherited gene for flower color.

Markers may Be:

- **Morphological** - These markers are often detectable by eye, by simple visual inspection. Examples of this type of marker include the presence or absence of an awn, leaf sheath coloration, height, grain color, aroma of rice etc. In well-characterized crops like maize, tomato, pea, barley or wheat, tens or hundreds of genes that determine morphological traits have been mapped to specific chromosome locations.
- **Biochemical-** A protein that can be extracted and observed; for example, isozymes and storage proteins.
- **Cytological** - The chromosomal banding produced by different stains; for example, G banding.
- **DNA-based**- Including microsatellites (also known as short tandem repeats, STRs, or simple sequence repeats, SSRs), restriction fragment length polymorphism (RFLP), random amplification of polymorphic DNA (RAPD), amplified fragment length polymorphism (AFlP), and single nucleotide polymorphisms (SNPs).

Positive and negative selectable markers

The following terms are generally less relevant to discussions of MAS in plant and animal breeding, but are highly relevant in molecular biology research:

- Positive selectable markers are selectable markers that confer selective advantage to the host organism. An example would be antibiotic resistance, which allows the host organism to survive antibiotic selection.
- Negative selectable markers are selectable markers that eliminate or inhibit growth of the host organism upon selection. An example would be thymidine kinase, which makes the host sensitive to ganciclovir selection.

A distinction can be made between selectable markers (which eliminate certain genotypes from the population) and screenable markers (which cause certain genotypes to be readily identifiable, at which point the experimenter must "score" or evaluate the population and act to retain the preferred genotypes). Most MAS uses screenable markers rather than selectable markers.

Gene vs Marker

The gene of interest directly causes production of protein(s) or RNA that produce a desired trait or phenotype, whereas markers (a DNA sequence or the morphological or biochemical markers produced due to that DNA) are genetically linked to the gene of interest. The gene of interest and the marker tend to move together during segregation of gametes due to their proximity on the same chromosome and concomitant reduction in recombination (chromosome crossover events) between the marker and gene of interest. For some traits, the gene of interest has been discovered and the presence of desirable alleles can be directly assayed with a high level of confidence. However, if the gene of interest is not known, markers linked to the gene of interest can still be used to select for individuals with desirable alleles of the gene of interest. When markers are used there may be some inaccurate results due to inaccurate tests for the marker. There also can be false positive results when markers are used, due to recombination between the marker of interest and gene (or QTL). A perfect marker would elicit no false positive results. The term 'perfect marker' is sometimes used when tests are performed to detect a SNP or other DNA polymorphism in the gene of interest, if that SNP or other polymorphism is the direct cause of the trait of interest. The term 'marker' is still appropriate to use when directly assaying the gene of interest, because the test of genotype is an indirect test of the trait or phenotype of interest.

Important properties of Ideal Markers for MAS

An ideal marker:

- Easy recognition of all possible phenotypes (homo- and heterozygotes) from all different alleles
- Demonstrates measurable differences in expression between trait types or gene of interest alleles, early in the development of the organism
- Testing for the marker does not have variable success depending on the allele at the marker locus or the allele at the target locus (the gene of interest that determines the trait of interest).
- Low or null interaction among the markers allowing the use of many at the same time in a segregating population
- Abundant in number
- Polymorphic

Drawbacks of Morphological Markers

Morphological markers are associated with several general deficits that reduce their usefulness including:

- the delay of marker expression until late into the development of the organism
- dominance
- deleterious effects
- pleiotropy
- confounding effects of genes unrelated to the gene or trait of interest but which also affect the morphological marker (epistasis)
- rare polymorphism
- frequent confounding effects of environmental factors which affect the morphological characteristics of the organism

To avoid problems specific to morphological markers, DNA-based markers have been developed. They are highly polymorphic, exhibit simple inheritance (often codominant), are abundant throughout the genome, are easy and fast to detect, exhibit minimum pleiotropic effects, and detection is not dependent on the developmental stage of the organism. Numerous markers have been mapped to different chromosomes in several crops including rice, wheat, maize, soybean and several others, and in livestock such as cattle, pigs and chickens. Those markers have been used in diversity analysis, parentage detection, DNA fingerprinting, and prediction of hybrid performance. Molecular markers are useful in indirect selection processes, enabling manual selection of individuals for further propagation.

SELECTION FOR MAJOR GENES LINKED TO MARKERS

'Major genes' that are responsible for economically important characteristics are frequent in the plant kingdom. Such characteristics include disease resistance, male sterility, self-incompatibility, and others related to shape, color, and architecture of whole plants and are often of mono- or oligogenic in nature. The marker loci that are tightly linked to major genes can be used for selection and are sometimes more efficient than direct selection for the target gene. Such advantages in efficiency may be due for example, to higher expression of the marker mRNA in such cases that the marker is itself a gene. Alternatively, in such cases that the target gene of interest differs between two alleles by a difficult-to-detect single nucleotide polymorphism, an external marker (be it another gene or a polymorphism that is easier to detect, such as a short tandem repeat) may present as the most realistic option.

Situations that are Favorable for Molecular Marker Selection

There are several indications for the use of molecular markers in the selection of a genetic trait.

In such Situations that:

- The selected character is expressed late in plant development, like fruit and flower features or adult characters with a juvenile period (so that it is not necessary to wait for the organism to become fully developed before arrangements can be made for propagation)
- The expression of the target gene is recessive (so that individuals which are heterozygous positive for the recessive allele can be crossed to produce some homozygous offspring with the desired trait)
- There is requirement for the presence of special conditions in order to invoke expression of the target gene(s), as in the case of breeding for disease and pest resistance (where inoculation with the disease or subjection to pests would otherwise be required). This advantage derives from the errors due to unreliable inoculation methods and the fact that field inoculation with the pathogen is not allowed in many areas for safety reasons. Moreover, problems in the recognition of the environmentally unstable genes can be eluded.
- The phenotype is affected by two or more unlinked genes (epistatis). For example, selection for multiple genes which provide resistance against diseases or insect pests for gene pyramiding.

The cost of genotyping (an example of a molecular marker assay) is reducing while the cost of phenotyping is increasing particularly in developed countries thus increasing the attractiveness of MAS as the development of the technology continues.

Steps for MAS

Generally the first step is to map the gene or quantitative trait locus (QTL) of interest first by using different techniques and then using this information for marker assisted selection. Generally, the markers to be used should be close to gene of interest (<5 recombination unit or cM) in order to ensure that only minor fraction of the selected individuals will be recombinants. Generally, not only a single marker but rather two markers are used in order to reduce the chances of an error due to homologous recombination. For example, if two flanking markers are used at same time with an interval between them of approximately 20cM, there is higher probability (99%) for recovery of the target gene.

QTL mapping techniques

In plants QTL mapping is generally achieved using bi-parental cross populations; a cross between two parents which have a contrasting phenotype for the trait of interest are developed. Commonly used populations are near isogenic lines (NILs), recombinant inbred lines (RILs), doubled haploids (DH), back cross and F2. Linkage between the phenotype and markers which have already been mapped is tested in these populations in order to determine the position of the QTL. Such techniques are based on linkage and are therefore referred to as "linkage mapping".

Single step MAS and QTL mapping

In contrast to two-step QTL mapping and MAS, a single-step method for breeding typical plant populations has been developed. In such an approach, in the first few breeding cycles, markers linked to the trait of interest are identified by QTL mapping and later the same information is used in the same population. In this approach, pedigree structure is created from families that are created by crossing number of parents (in three-way or four way crosses). Both phenotyping and genotyping is done using molecular markers mapped the possible location of QTL of interest. This will identify markers and their favorable alleles. Once these favorable marker alleles are identified, the frequency of such alleles will be increased and response to marker assisted selection is estimated. Marker allele(s) with desirable effect will be further used in next selection cycle or other experiments.

High-throughput genotyping techniques

Recently high-throughput genotyping techniques are developed which allows marker aided screening of many genotypes. This will help breeders in shifting traditional breeding to marker aided selection. One example of such automation is using DNA isolation robots, capillary electrophoresis and pipetting robots.

One recent example of capllilary system is Applied Biosystems 3130 Genetic Analyzer. This is the latest generation of 4-capillary electrophoresis instruments for the low to medium throughput laboratories.

Use of MAS for Backcross Breeding

A minimum of five or six-backcross generations are required to transfer a gene of interest from a donor (may not be adapted) to a recipient (recurrent – adapted cultivar). The recovery of the recurrent genotype can be accelerated with the use of molecular markers. If the F1 is heterozygous for the marker locus, individuals with the recurrent parent allele(s) at the marker locus in first or subsequent backcross generations will also carry a chromosome tagged by the marker.

Marker assisted gene pyramiding

Gene pyramiding has been proposed and applied to enhance resistance to disease and insects by selecting for two or more than two genes at a time. For example, in rice such pyramids have been developed against bacterial blight and blast. The advantage of use of markers in this case allows to select for QTL-allele-linked markers that have same phenotypic effect.

A coordinated effort to implement wheat (Triticum turgidum and Triticum aestivum) marker assisted selection in the U.S. as well as a resource for marker assisted selection exists at the Wheat CAP (Coordinated Agricultural Project) website.

Reverse breeding and Doubled Haploids (DH)

Homozygous plants with desirable traits can be produced from heterozygous starting plants, if a haploid cell with the alleles for those traits can be produced, and then used to make a doubled haploid. The doubled haploid will be homozygous for the desired traits. Furthermore, two different homozygous plants created in that way can be used to produce a generation of F1 hybrid plants which have the advantages of heterozygosity and a greater range of possible traits. Thus, an individual heterozygous plant chosen for its desirable characteristics can be converted into a heterozygous variety (F1 hybrid) without the necessity of vegetative reproduction but as the result of the cross of two homozygous/doubled haploid lines derived from the originally selected plant. Using plant tissue culturing can produce haploid or double haploid plant lines and generations. This minimizes the amount of genetic diversity among that plant species in order to select for desirable traits that will increase the fitness of the individuals. Using this method decreases the need for breeding multiple generations of plants to get a generation that is homologous for the desired traits, therefore save much time in the process. There are many plant tissue culturing techniques that can be used to achieve the haploid plants, but microspore culturing is currently the most promising for producing the largest numbers of them.A doubled haploid (DH) is a genotype formed when haploid cells undergo chromosome doubling. Artificial production of doubled haploids is important in plant breeding. Haploid cells are produced from pollen or egg cells or from other cells of the gametophyte, then by induced or spontaneous chromosome doubling, a doubled haploid cell is produced, which can be grown into a doubled haploid plant. If the original plant was diploid, the haploid cells are monoploid, and the term doubled monoploid may be used for the doubled haploids. Haploid organisms derived from tetraploids or hexaploids are sometimes called dihaploids (and the doubled dihaploids are, respectively, tetraploid or hexaploid). Conventional inbreeding procedures take six generations to achieve approximately complete homozygosity, whereas doubled haploidy achieves it in one generation. Dihaploid plants derived from tetraploid crop plants may be important for breeding programs that involve diploid wild relatives of the crops.

History

The first report of the haploid plant was published by Blakeslee et al. (1922) in Datura stramonium. Subsequently, haploids were reported in many other species. Guha and Maheshwari (1964) developed an anther culture technique for the production of haploids in the laboratory. Haploid production by wide crossing was reported in barley (Kasha and Kao, 1970) and tobacco (Burk et al., 1979). Tobacco, rapeseed, and barley are the most responsive species for doubled haploid production. Doubled haploid methodologies have now been applied to over 250 species.

PRODUCTION OF DOUBLED HAPLOIDS

Doubled haploids can be produced in vivo or in vitro. Haploid embryos are produced in vivo by parthenogenesis, pseudogamy, or chromosome elimination after wide crossing. The haploid embryo is rescued, cultured, and chromosome-doubling produces doubled haploids. The in vitro methods include gynogenesis (ovary and flower culture) and androgenesis (anther and microspore culture). Androgenesis is the preferred method. Another method of producing the haploids is wide crossing. In barley, haploids can be produced by wide crossing with the related species Hordeum bulbosum; fertilization is affected, but during the early stages of seed development the H. bulbosum chromosomes are eliminated leaving a haploid embryo. In tobacco (Nicotiana tabacum), wide crossing with Nicotiana africana is widely used. When N. africana is used to pollinate N. tabacum, 0.25 to 1.42 percent of the progenysurvive and can readily be identified as either F1 hybrids or maternal haploids.

Although these percentages appear small, the vast yield of tiny seeds and the early death of most seedlings provide significant numbers of viable hybrids and haploids in relatively small soil containers. This method of interspecific pollination serves as a practical way of producing seed-derived haploids of N. tabacum, either as an alternative method or complementary method to anther culture.

Genetics of DH Population

In DH method only two types of genotypes occur for a pair of alleles, A and a, with the frequency of ½ AA and ½ aa, while in diploid method three genotypes occur with the frequency of ¼ AA, ½ Aa, ¼ aa. Thus, if AA is desirable genotype, the probability of obtaining this genotype is higher in haploid method than in diploid method. If n loci are segregating, the probability of getting the desirable genotype is (1/2)n by the haploid method and (1/4)n by the diploid method. Hence the efficiency of the haploid method is high when the number of genes concerned is large.

Studies were conducted comparing DH method and other conventional breeding methods and it was concluded that adoption of doubled haploidy does not lead to any bias of genotypes in populations, and random DHs were even found to be compatible to selected line produced by conventional pedigree method.

APPLICATIONS OF DHS PLANT BREEDING

Mapping Quantitative Trait Loci

Most of the economic traits are controlled by genes with small but cumulative effects. Although the potential of DH populations in quantitative genetics has been understood for some time, it was the advent of molecular marker maps that provided the impetus for their use in identifying loci controlling quantitative traits. As the quantitative trait loci (QTL) effects are small and highly influenced by environmental factors, accurate phenotyping with replicated trials is needed. This is possible with doubled haploidy organisms because of their true breeding nature and because they can conveniently be produced in large numbers. Using DH populations, 130 quantitative traits have been mapped in nine crop species. In total, 56 DH populations were used for QTL detection.

Backcross breeding

In backcross conversion, genes are introgressed from a donor cultivar or related species into a recipient elite line through repeated backcrossing. A problem in this procedure is being able to identify the lines carrying the trait of interest at each generation. The problem is particularly acute if the trait of interest is recessive, as it will be present only in a heterozygous condition after each backcross. The development of molecular markers provides an easier method of selection based on the genotype (marker) rather than the phenotype. Combined with doubled haploidy it becomes more effective. In marker assisted backcross conversion, a recipient parent is crossed with a donor line and the hybrid (F1) backcrossed to the recipient. The resulting generation (BC1) is backcrossed and the process repeated until the desired genotypes are produced. The combination of doubled haploidy and molecular marker provides the short cut. In the back cross generation one itself a genotype with the character of interest can be selected and converted into homozygous doubled haploid genotype. Chen et al. (1994) used marker assisted backcross conversion with doubled haploidy of BC1 individuals to select stripe rust resistant lines in barley.

Bulked Segregant Analysis (BSA)

In bulked segregant analysis, a population is screened for a trait of interest and the genotypes at the two extreme ends form two bulks. Then the two bulks are tested for the presence or absence of molecular markers. Since the bulks are supposed to contrast in the alleles that contribute positive and negative effects, any marker polymorphism between the two bulks indicates the linkage between the marker and trait of interest. BSA is dependent on accurate phenotyping and the DH population has particular advantage in that they are true breeding and can be tested repeatedly. DH populations are commonly used in bulked segregant analysis, which is a popular method in marker assisted breeding. This method has been applied mostly to rapeseed and barley.

Genetic maps

Genetic maps are very important to understand the structure and organization of genomes from which evolution patterns and syntenic relationships between species can be deduced. Genetic maps also provide a framework for the mapping of genes of interest and estimating the magnitude of their effects and aid our understanding of genotype/phenotype associations. DH populations have become standard resources in genetic mapping for species in which DHs are readily available. Doubled haploid populations are ideal for genetic mapping. It is possible to produce a genetic map within two years of the initial cross regardless of the species. Map construction is relatively easy using a DH population derived from a hybrid of two homozygous parents as the expected segregation ratio is simple, i.e. 1:1. DH populations have now been used to produce genetic maps of barley, rapeseed, rice, wheat, and pepper. DH populations played a major role in facilitating the generation of the molecular marker maps in eight crop species.

Genetic Studies

Genetic ratios and mutation rates can be read directly from haploid populations. A small doubled haploid (DH) population was used to demonstrate that a dwarfing gene in barley is located chromosome 5H. In another study the segregation of a range of markers has been analyzed in barley.

Genomics

Although QTL analysis has generated a vast amount of information on gene locations and the magnitude of effects on many traits, the identification of the genes involved has remained elusive. This is due to poor resolution of QTL analysis. The solution for this problem would be production of recombinant chromosome substitution line, or stepped aligned recombinant inbred lines. Here, backcrossing is carried out until a desired level of recombination has occurred and genetic markers are used to detect desired recombinant chromosome substitution lines in the target region, which can be fixed by doubled haploidy. In rice, molecular markers have been found to be linked with major genes and QTLs for resistance to rice blast, bacterial blight, and sheath blight in a map produced from DH population.

Elite Crossing

Traditional breeding methods are slow and take 10–15 years for cultivar development. Another disadvantage is inefficiency of selection in early generations because of heterozygosity. These two disadvantages can be over come by DHs, and more elite crosses can be evaluated and selected within less time.

Cultivar development

Uniformity is a general requirement of cultivated line in most species, which can be easily obtained through DH production. There are various ways in which DHs can be used in cultivar production. The DH lines themselves can be released as cultivars, they may be used as parents in hybrid cultivar production or more indirectly in the creation of breeders lines and in germplasm conservation. Barley has over 100 direct DH cultivars. According to published information there are currently around 300 DH derived cultivars in 12 species worldwide.

The relevance of DHs to plant breeding has increased markedly in recent years owing to the development of protocols for 25 species. Doubled haploidy already plays an important role in hybrid cultivar production of vegetables, and the potential for ornamental production is being vigorously examined. DHs are also being developed in the medicinal herb Valeriana officinalis to select lines with high pharmacological activity. Another interesting development is that fertile homozygous DH lines can be produced in species that have self-incompatibility systems.

Advantages of DHs

The ability to produce homozygous lines after a single round recombination saves a lot of time for the plant breeders. Studies conclude that random DH's are comparable to the selected lines in pedigree inbreeding. The other advantages include development of large number of homozygous lines, efficient genetic analysis and development of markers for useful traits in much less time. More specific benefits include the possibility of seed propagation as an alternative to vegetative multiplication in ornamentals, and in species such as trees in which long life cycles and inbreeding depression preclude traditional breeding methods, doubled haploidy provides new alternatives.

Disadvantages of DHs

The main disadvantage with the DH population is that selection cannot be imposed on the population. But in conventional breeding selection can be practised for several generations: thereby desirable characters can be improved in the population. In haploids produced from anther culture, it is observed that some plants are aneuploids and some are mixed haploid-diploid types. Another disadvantage associated with the double haploidy is the cost involved in establishing tissue culture and growth facilities. The over-usage of doubled haploidy may reduce genetic variation in breeding germplasm. Hence one has to take several factors into consideration before deploying doubled haploidy in breeding programmes.

Conclusions

Technological advances have now provided DH protocols for most plant genera. The number of species amenable to doubled haploidy has reached a staggering 250 in just a few decades. Response efficiency has also improved with gradual removal of species from recalcitrant category. Hence it will provide greater efficiency of plant breeding.

Genetic modification

Genetic modification of plants is achieved by adding a specific gene or genes to a plant, or by knocking down a gene with RNAi, to produce a desirable phenotype. The plants resulting from adding a gene are often referred to as transgenic plants. If for genetic modification genes of the species or of a crossable plant are used under control of their native promoter, then they are called cisgenic plants. Sometimes genetic modification can produce a plant with the desired trait or traits faster than classical breeding because the majority of the plant's genome is not altered.

To genetically modify a plant, a genetic construct must be designed so that the gene to be added or removed will be expressed by the plant. To do this, a promoter to drive transcription and a termination sequence to stop transcription of the new gene, and the gene or genes of interest must be introduced to the plant. A marker for the selection of transformed plants is also included. In the laboratory, antibiotic resistance is a commonly used marker: Plants that have been successfully transformed will grow on media containing antibiotics; plants that have not been transformed will die. In some instances markers for selection are removed by backcrossing with the parent plant prior to commercial release.

The construct can be inserted in the plant genome by genetic recombination using the bacteria Agrobacterium tumefaciens or A. rhizogenes, or by direct methods like the gene gun or microinjection. Using plant viruses to insert genetic constructs into plants is also a possibility, but the technique is limited by the host range of the virus. For example, Cauliflower mosaic virus (CaMV) only infects cauliflower and related species. Another limitation of viral vectors is that the virus is not usually passed on the progeny, so every plant has to be inoculated.

The majority of commercially released transgenic plants are currently limited to plants that have introduced resistance to insectpests and herbicides. Insect resistance is achieved through incorporation of a gene from Bacillus thuringiensis (Bt) that encodes a protein that is toxic to some insects. For example, the cotton bollworm, a common cotton pest, feeds on Bt cotton it will ingest the toxin and die. Herbicides usually work by binding to certain plant enzymes and inhibiting their action. The enzymes that the herbicide inhibits are known as the herbicides target site. Herbicide resistance can be engineered into crops by expressing a version of target site protein that is not inhibited by the herbicide. This is the method used to produce glyphosate resistant crop plants (See Glyphosate)

Genetic modification can further increase yields by increasing stress tolerance to a given environment. Stresses such as temperature variation, are signalled to the plant via a cascade of signalling molecules which will activate a Transcription factor to regulate Gene expression. Overexpression of particular genes involved in cold acclimation has been shown to become more resistant to freezing, which is one common cause of yield loss Genetic modification of plants that can produce pharmaceuticals (and industrial chemicals), sometimes called pharming, is a rather radical new area of plant breeding. Much of the advances in the field genetic engineering has come from experimentation with tobacco. Major advances in tissue culture and plant cellular mechanisms for a wide range of plants has originated from systems developed in tobacco. It was the first plant to be genetically engineered and is considered a model organism for not only genetic engineering, but a range of other fields. As such the transgenic tools and procedures are well established making it one of the easiest plants to transform. Another major model organism relevant to genetic engineering is Arabidopsis thaliana. Its small genome and short life cycle makes it easy to manipulate and it contains many homolgs to important crop species. It was the first plant sequenced, has abundant bioinformatic resources and can be transformed by simply dipping a flower in a transformed Agrobacterium solution.

In research, plants are engineered to help discover the functions of certain genes. The simplest way to do this is to remove the gene and see what phenotype develops compared to the wild type form. Any differences are possibly the result of the missing gene. Unlike mutagenisis, genetic engineering allows targeted removal without disrupting other genes in the organism. Some genes are only expressed in certain tissue, so reporter genes, like GUS, can be attached to the gene of interest allowing visualisation of the location. Other ways to test a gene is to alter it slightly and then return it to the plant and see if it still has the same effect on phenotype. Other strategies include attaching the gene to a strong promoter and see what happens when it is over expressed, forcing a gene to be expressed in a different location or at different developmental stages.

Ornamental

Some genetically modified plants are purely ornamental. They are modified for lower color, fragrance, flower shape and plant architecture. The first genetically modified ornamentals commercialised altered colour. Carnations were released in 1997, with the most popular genetically modified organism, a blue rose (actually lavender or mauve) created in 2004. The roses are sold in Japan, the United States, and Canada. Other genetically modified ornamentals include Chrysanthemum and Petunia. As well as increasing aesthetic value there are plans to develop ornamentals that use less water or are resistant to the cold, which would allow them to be grown outside their natural environments.

Conservation

It has been proposed to genetically modify some plant species threatened by extinction to be resistant invasive plants and diseases, such as the emerald ash borer in North American and the fungal disease, Ceratocystis platani, in European plane trees. The papaya ringspot virus (PRSV) devastated papaya trees in Hawaii in the twentieth century until transgenic papaya plants were given pathogen-derived resistance. However, genetic modification for conservation in plants remains mainly speculative. A unique concern is that a transgenic species may no longer bear enough resemblance to the original species to truly claim that the original species is being conserved. Instead, the transgenic species may be genetically different enough to be considered a new species, thus diminishing the conservation worth of genetic modification.

Crops

Genetically modified crops are genetically modified plants that are used in agriculture. The first crops provided are used for animal or human food and provide resistance to certain pests, diseases, environmental conditions, spoilage or chemical treatments (e.g. resistance to aherbicide). The second generation of crops aimed to improve the quality, often by altering the nutrient profile. Third generation genetically modified crops can be used for non-food purposes, including the production of pharmaceutical agents, biofuels, and other industrially useful goods, as well as for bioremediation.

There are three main aims to agricultural advancement; increased production, improved conditions for agricultural workers and sustainability. GM crops contribute by improving harvests through reducing insect pressure, increasing nutrient value and tolerating different abiotic stresses. Despite this potential, as of 2018, the commercialised crops are limited mostly to cash crops like cotton, soybean, maize and canola and the vast majority of the introduced traits provide either herbicide tolerance or insect resistance. Soybeans accounted for half of all genetically modified crops planted in 2014. Adoption by farmers has been rapid, between 1996 and 2013, the total surface area of land cultivated with GM crops increased by a factor of 100, from 17,000 square kilometers (4,200,000 acres) to 1,750,000 km (432 million acres). Geographically though the spread has been very uneven, with strong growth in the Americas and parts of Asia and little in Europe and Africa. Its socioeconomic spread has been more even, with approximately 54% of worldwide GM crops grown in developing countries in 2013.

Food

The majority of GM crops have been modified to be resistant to selected herbicides, usually a glyphosate or glufosinate based one. Genetically modified crops engineered to resist herbicides are now more available than conventionally bred resistant varieties; in the USA 93% of soybeans and most of the GM maize grown is glyphosate tolerant. Most currently available genes used to engineer insect resistance come from the Bacillus thuringiensis bacterium. Most are in the form of delta endotoxin genes known as cry proteins, while a few use the genes that encode for vegetative insecticidal proteins. The only gene commercially used to provide insect protection that does not originate from *B. thuringiensis* is the Cowpeatrypsin inhibitor (CpTI). CpTI was first approved for use cotton in 1999 and is currently undergoing trials in rice. Less than one percent of GM crops contained other traits, which include providing virus resistance, delaying senescence, modifying flower colour and altering the plants composition. Golden rice is the most well known GM crop that is aimed at increasing nutrient value. It has been engineered with three genes that biosynthesisebeta-carotene, a precursor of vitamin A, in the edible parts of rice. It is intended to produce a fortified food to be grown and consumed in areas with a shortage of dietary vitamin A.a deficiency which each year is estimated to kill 670,000 children under the age of 5 and cause an additional 500,000 cases of irreversible childhood blindness. The original golden rice produced 1.6ìg/g of the carotenoids, with further development increasing this 23 times. In 2018 it gained its first approvals for use as food.

Issues and concerns

Modern plant breeding, whether classical or through genetic engineering, comes with issues of concern, particularly with regard to food crops. The question of whether breeding can have a negative effect on nutritional value is central in this respect. Although relatively little direct research in this area has been done, there are scientific indications that, by favoring certain aspects of a plant's development, other aspects may be retarded. A study published in the Journal of the American College of Nutrition in 2004, entitled Changes in USDA Food Composition Data for 43 Garden Crops, 1950 to 1999, compared nutritional analysis of vegetables done in 1950 and in 1999, and found substantial decreases in six of 13 nutrients measured, including 6% of protein and 38% of riboflavin. Reductions in calcium, phosphorus, iron and ascorbic acid were also found. The study, conducted at the Biochemical Institute, University of Texas at Austin, concluded in summary: "We suggest that any real declines are generally most easily explained by changes in cultivated varieties between 1950 and 1999, in which there may be trade-offs between yield and nutrient content."

The debate surrounding genetically modified food during the 1990s peaked in 1999 in terms of media coverage and risk perception, and continues today – for example, "Germany has thrown its weight behind a growing European mutiny over genetically modified crops by banning the planting of a widely grown pest-resistant corn variety." The debate encompasses the ecological impact of genetically modified plants, the safety of genetically modified food and concepts used for safety evaluation like substantial equivalence. Such concerns are not new to plant breeding. Most countries have regulatory processes in place to help ensure that new crop varieties entering the marketplace are both safe and meet farmers' needs.

Examples include variety registration, seed schemes, regulatory authorizations for GM plants, etc.

Plant breeders' rights is also a major and controversial issue. Today, production of new varieties is dominated by commercial plant breeders, who seek to protect their work and collect royalties through national and international agreements based in intellectual property rights. The range of related issues is complex. In the simplest terms, critics of the increasingly restrictive regulations argue that, through a combination of technical and economic pressures, commercial breeders are reducing biodiversity and significantly constraining individuals (such as farmers) from developing and trading seed on a regional level. Efforts to strengthen breeders' rights, for example, by lengthening periods of variety protection, are ongoing.

When new plant breeds or cultivars are bred, they must be maintained and propagated. Some plants are propagated by asexual means while others are propagated by seeds. Seed propagated cultivars require specific control over seed source and production procedures to maintain the integrity of the plant breeds results. Isolation is necessary to prevent cross contamination with related plants or the mixing of seeds after harvesting. Isolation is normally accomplished by planting distance but in certain crops, plants are enclosed in

Greenhouses or Cages (most commonly used when producing F1 hybrids.)

Organic farming is an alternative agricultural system which originated early in the 20th century in reaction to rapidly changing farming practices. Organic farming continues to be developed by various organic agriculture organizations today,it also defined as use of relies on fertilizers of organic origin such as compost manure, green manure, and bone meal and places emphasis on techniques such as crop rotation and companion planting. Biological pest control, mixed cropping and the fostering of insect predators are encouraged. In general, organic standards are designed to allow the use of naturally occurring substances while prohibiting or strictly limiting synthetic substances. For instance, naturally occurring pesticides such as pyrethrin and rotenone are permitted, while synthetic fertilizers and pesticides are generally prohibited. Synthetic substances that are allowed include, for example, copper sulfate, elemental sulfur and Ivermectin. Genetically modified organisms, nanomaterials, human sewage sludge, plant growth regulators, hormones, and antibiotic use in livestock husbandry are prohibited. Reasons for advocation of organic farming include advantages in sustainability,openness, self-sufficiency, autonomy/independence,health, food security, and food safety.

Organic agricultural methods are internationally regulated and legally enforced by many nations, based in large part on the standards set by the International Federation of Organic Agriculture Movements (IFOAM), an international umbrella organization for organic farming organizations established in 1972. Organic agriculture can be defined as:

An integrated farming system that strives for sustainability, the enhancement of soil fertility and biological diversity whilst, with rare exceptions, prohibiting synthetic pesticides, antibiotics, synthetic fertilizers, genetically modified organisms, and growth hormones. Since 1990 the market for organic food and other products has grown rapidly, reaching $63 billion worldwide in 2012.This demand has driven a similar increase in organically managed farmland that grew from 2001 to 2011 at a compounding rate of 8.9% per annum. As of 2019, approximately 70,000,000 hectares (170,000,000 acres) worldwide were farmed organically, representing approximately 1.4 percent of total world farmland.

Role of Plant breeding in Organic Agriculture

Critics of organic agriculture claim it is too low-yielding to be a viable alternative to conventional agriculture. However, part of that poor performance may be the result of growing poorly adapted varieties. It is estimated that over 95% of organic agriculture is based on conventionally adapted varieties, even though the production environments found in organic vs. conventional farming systems are vastly different due to their distinctive management practices. Most notably, organic farmers have fewer inputs available than conventional growers to control their production environments. Breeding varieties specifically adapted to the unique conditions of organic agriculture is critical for this sector to realize its full potential. This requires selection for traits such as:

- Water use efficiency
- Nutrient use efficiency (particularly nitrogen and phosphorus)
- Weed competitiveness
- Tolerance of mechanical weed control
- Pest/disease resistance
- Early maturity (as a mechanism for avoidance of particular stresses)
- Abiotic stress tolerance (i.e. drought, salinity, etc...)

Currently, few breeding programs are directed at organic agriculture and until recently those that did address this sector have generally relied on indirect selection (i.e. selection in conventional environments for traits considered important for organic agriculture). However, because the difference between organic and conventional environments is large, a given genotype may perform very differently in each environment due to an interaction between genes and the environment (see gene-environment interaction). If this interaction is severe enough, an important trait required for the organic environment may not be revealed in the conventional environment, which can result in the selection of poorly adapted individuals. To ensure the most adapted varieties are identified, advocates of organic breeding now promote the use of direct selection (i.e. selection in the target environment) for many agronomic traits.

There are many classical and modern breeding techniques that can be utilized for crop improvement in organic agriculture despite the ban on genetically modified organisms. For instance, controlled crosses between individuals allow desirable genetic variation to be recombined and transferred to seed progeny via natural processes. Marker assisted selection can also be employed as a diagnostics tool to facilitate selection of progeny who possess the desired trait(s), greatly speeding up the breeding process. This technique has proven particularly useful for the introgression of resistance genes into new backgrounds, as well as the efficient selection of many resistance genes pyramided into a single individual. Unfortunately, molecular markers are not currently available for many important traits, especially complex ones controlled by many genes.

Addressing global food security through plant breeding

For future agriculture to thrive there are necessary changes which must be made in accordance to arising global issues. These issues are arable land, harsh cropping conditions and food security which involves, being able to provide the world population with food containing sufficient nutrients. These crops need to be able to mature in several environments allowing for worldwide access, this is involves issues such as drought tolerance. These global issues are achievable through the process of plant breeding, as it offers the ability to select specific genes allowing the crop to perform at a level which yields the desired results.

Increased yield without expansion

With an increasing population, the production of food needs to increase with it. It is estimated that a 70% increase in food production is needed by 2050 in order to meet the Declaration of the World Summit on Food Security. But with the degradation of agricultural land, simply planting more crops is no longer a viable option. New varieties of plants can in some cases be developed through plant breeding that generate an increase of yield without relying on an increase in land area. An example of this can be seen in Asia, where food production per capita has increased twofold. This has been achieved through not only the use of fertilisers, but through the use of better crops that have been specifically designed for the area.

Breeding for Increased Nutritional Value

Plant breeding can contribute to global food security as it is a cost-effective tool for increasing nutritional value of forage and crops. Improvements in nutritional value for forage crops from the use of analytical chemistry and rumen fermentation technology have been recorded since 1960; this science and technology gave breeders the ability to screen thousands of samples within a small amount of time, meaning breeders could identify a high performing hybrid quicker. The main area genetic increases were made was in vitro dry matter digestibility (IVDMD) resulting in 0.7-2.5% increase, at just 1% increase in IVDMD a single Bos Taurus also known as beef cattle reported 3.2% increase in daily gains. This improvement indicates plant breeding is an essential tool in gearing future agriculture to perform at a more advanced level.

Breeding for tolerance

Plant breeding of hybrid crops has become extremely popular worldwide in an effort to combat the harsh environment. With long periods of drought and lack of water or nitrogen stress tolerance has become a significant part of agriculture. Plant breeders have focused on identifying crops which will ensure crops perform under these conditions; a way to achieve this is finding strains of the crop that is resistance to drought conditions with low nitrogen. It is evident from this that plant breeding is vital for future agriculture to survive as it enables farmers to produce stress resistant crops hence improving food security. In countries that experience harsh winters such as Iceland, Germany and further east in Europe, plant breeders are involved in breeding for tolerance to frost, continuous snow-cover, frost-drought (desiccation from wind and solar radiation under frost) and high moisture levels in soil in winter.

Participatory plant breeding

Participatory plant breeding (PPB) is when farmers are involved in a crop improvement programme with opportunities to make decisions and contribute to the research process at different stages. Participatory approaches to crop improvement can also be applied when plant biotechnologies are being used for crop improvement. Local agricultural systems and genetic diversity are developed and strengthened by crop improvement, which participatory crop improvement (PCI) plays a large role. PPB is enhanced by farmers knowledge of the quality required and evaluation of target environment which affects the effectiveness of PPB.

Organic farming is an alternative agricultural system which originated early in the 20th century in reaction to rapidly changing farming practices. Organic farming continues to be developed by various organic agriculture organizations today,it also defined as use of relies on fertilizers of organic origin such as compost manure, green manure, and bone meal and places emphasis on techniques such as crop rotation and companion planting. Biological pest control, mixed cropping and the fostering of insect predators are encouraged. In general, organic standards are designed to allow the use of naturally occurring substances while prohibiting or strictly limiting synthetic substances. For instance, naturally occurring pesticides such as pyrethrin and rotenone are permitted, while synthetic fertilizers and pesticides are generally prohibited. Synthetic substances that are allowed include, for example, copper sulfate, elemental sulfur and Ivermectin. Genetically modified organisms, nanomaterials, human sewage sludge, plant growth regulators, hormones, and antibiotic use in livestock husbandry are prohibited. Reasons for advocation of organic farming include advantages in sustainability,openness, self-sufficiency, autonomy/independence,health, food security, and food safety.

Organic agricultural methods are internationally regulated and legally enforced by many nations, based in large part on the standards set by the International Federation of Organic Agriculture Movements (IFOAM), an international umbrella organization for organic farming organizations established in 1972. Organic agriculture can be defined as: an integrated farming system that strives for sustainability, the enhancement of soil fertility and biological diversity whilst, with rare exceptions, prohibiting synthetic pesticides, antibiotics, synthetic fertilizers, genetically modified organisms, and growth hormones.Since 1990 the market for organic food and other products has grown rapidly, reaching $63 billion worldwide in 2012.This demand has driven a similar increase in organically managed farmland that grew from 2001 to 2011 at a compounding rate of 8.9% per annum. As of 2019, approximately 70,000,000 hectares (170,000,000 acres) worldwide were farmed organically, representing approximately 1.4 percent of total world farmland.

Methods

Organic agriculture is a production system that sustains the health of soils, ecosystems and people. It relies on ecological processes, biodiversity and cycles adapted to local conditions, rather than the use of inputs with adverse effects. Organic agriculture combines tradition, innovation and science to benefit the shared environment and promote fair relationships and a good quality of life for all involved...

Organic farming methods combine scientific knowledge of ecology and some modern technology with traditional farming practices based on naturally occurring biological processes. Organic farming methods are studied in the field of agroecology. While conventional agriculture uses synthetic pesticides and water-soluble synthetically purified fertilizers, organic farmers are restricted by regulations to using natural pesticides and fertilizers. An example of a natural pesticide is pyrethrin, which is found naturally in the Chrysanthemum flower. The principal methods of organic farming include crop rotation, green manures and compost, biological pest control, and mechanical cultivation. These measures use the natural environment to enhance agricultural productivity: legumes are planted to fix nitrogen into the soil, natural insect predators are encouraged, crops are rotated to confuse pests and renew soil, and natural materials such as potassium bicarbonate and mulches are used to control disease and weeds. Genetically modified seeds and animals are excluded. While organic is fundamentally different from conventional because of the use of carbon based fertilizers compared with highly soluble synthetic based fertilizers and biological pest control instead of synthetic pesticides, organic farming and large-scale conventional farming are not entirely mutually exclusive. Many of the methods developed for organic agriculture have been borrowed by more conventional agriculture. For example, Integrated Pest Management is a multifaceted strategy that uses various organic methods of pest control whenever possible, but in conventional farming could include synthetic pesticides only as a last resort.

Crop diversity

Organic farming encourages Crop diversity. The science of agroecology has revealed the benefits of polyculture (multiple crops in the same space), which is often employed in organic farming. Planting a variety of vegetable crops supports a wider range of beneficial insects, soil microorganisms, and other factors that add up to overall farm health. Crop diversity helps environments thrive and protects species from going extinct.

Soil management

Organic farming relies heavily on the natural breakdown of organic matter, using techniques like green manure and composting, to replace nutrients taken from the soil by previous crops. This biological process, driven by microorganisms such as mycorrhiza, allows the natural production of nutrients in the soil throughout the growing season, and has been referred to as feeding the soil to feed the plant.

Organic farming uses a variety of methods to improve soil fertility, including crop rotation, cover cropping, reduced tillage, and application of compost. By reducing tillage, soil is not inverted and exposed to air; less carbon is lost to the atmosphere resulting in more soil organic carbon. This has an added benefit of carbon sequestration, which can reduce green house gases and help reverse climate change .

Plants need a large number of nutrients in various quantities to flourish. Supplying enough nitrogen and particularly synchronization so that plants get enough nitrogen at the when plants need it most, is a challenge for organic farmers.Crop rotation and green manure ("cover crops") help to provide nitrogen through legumes (more precisely, the Fabaceae family), which fix nitrogen from the atmosphere through symbiosis with rhizobialbacteria. Intercropping, which is sometimes used for insect and disease control, can also increase soil nutrients, but the competition between the legume and the crop can be problematic and wider spacing between crop rows is required. Crop residues can be ploughed back into the soil, and different plants leave different amounts of nitrogen, potentially aiding synchronization. Organic farmers also use animal manure, certain processed fertilizers such as seed meal and various mineral powders such as rock phosphate and green sand, a naturally occurring form of potash that provides potassium. Together these methods help to control erosion. In some cases pH may need to be amended. Natural pH amendments include lime and sulfur, but in the U.S. some compounds such as iron sulfate, aluminum sulfate, magnesium sulfate, and soluble boron products are allowed in organic farming.

Mixed farms with both livestock and crops can operate as ley farms, whereby the land gathers fertility through growing nitrogen-fixing forage grasses such as white clover or alfalfa and grows cash crops or cereals when fertility is established. Farms without livestock ("stockless") may find it more difficult to maintain soil fertility, and may rely more on external inputs such as imported manure as well as grain legumes and green manures, although grain legumes may fix limited nitrogen because they are harvested. Horticultural farms that grow fruits and vegetables in protected conditions often rely even more on external inputs.

Biological research into soil and soil organisms has proven beneficial to organic farming. Varieties of bacteria and fungi break down chemicals, plant matter and animal waste into productive soil nutrients. In turn, they produce benefits of healthier yields and more productive soil for future crops. Fields with less or no manure display significantly lower yields, due to decreased soil microbe community.

Increased manure improves biological activity, providing a healthier, more arable soil system and higher yields.

Weed Management

Organic weed management promotes weed suppression, rather than weed elimination, by enhancing crop competition and phytotoxic effects on weeds. Organic farmers integrate cultural, biological, mechanical, physical and chemical tactics to manage weeds without synthetic herbicides.

Organic standards require rotation of annual crops, meaning that a single crop cannot be grown in the same location without a different, intervening crop. Organic crop rotations frequently include weed-suppressive cover crops and crops with dissimilar life cycles to discourage weeds associated with a particular crop.Research is ongoing to develop organic methods to promote the growth of natural microorganisms that suppress the growth or germination of common weeds.

Other cultural practices used to enhance crop competitiveness and reduce weed pressure include selection of competitive crop varieties, high-density planting, tight row spacing, and late planting into warm soil to encourage rapid crop germination.

Mechanical and physical weed control practices used on organic farms can be broadly grouped as:

- Tillage - Turning the soil between crops to incorporate crop residues and soil amendments; remove existing weed growth and prepare a seedbed for planting; turning soil after seeding to kill weeds, including cultivation of row crops;
- and cutting - Removing top growth of weeds;
- Flame weeding and thermal weeding - Using heat to kill weeds; and
- Mulching - Blocking weed emergence with organic materials, plastic films, or landscape fabric. Some critics, citing work published in 1997 by David Pimentel of Cornell University, which described an epidemic of soil erosion worldwide, have raised concerned that tillage contribute to the erosion epidemic. The FAO and other organizations have advocated a 'no-till' approach to both conventional and organic farming, and point out in particular that crop rotation techniques used in organic farming are excellent no-till approaches. A study published in 2005 by Pimentel and colleagues confirmed that 'Crop rotations and cover cropping (green

manure) typical of organic agriculture reduce soil erosion, pest problems, and pesticide use.'

Some naturally sourced chemicals are allowed for herbicidal use. These include certain formulations of acetic acid (concentrated vinegar), corn gluten meal, and essential oils. A few selective bioherbicides based on fungal pathogens have also been developed. At this time, however, organic herbicides and bioherbicides play a minor role in the organic weed control toolbox.

Weeds can be controlled by grazing. For example, geese have been used successfully to weed a range of organic crops including cotton, strawberries, tobacco, and corn, reviving the practice of keeping cotton patch geese, common in the southern U.S. before the 1950s. Similarly, some rice farmers introduce ducks and fish to wet paddy fields to eat both weeds and insects.

Controlling other organisms

Organisms aside from weeds that cause problems on organic farms include arthropods (e.g., insects, mites), nematodes, fungi and bacteria. Organic practices include, but are not limited to:

- encouraging predatory beneficial insects to control pests by serving them nursery plants and/or an alternative habitat, usually in a form of a shelterbelt, hedgerow, or beetle bank;
- encouraging beneficial microorganisms;
- rotating crops to different locations from year to year to interrupt pest reproduction cycles;
- planting companion crops and pest-repelling plants that discourage or divert pests;
- row covers to protect crops during pest migration periods;
- using biologic pesticides and herbicides;
- using stale seed beds to germinate and destroy weeds before planting;
- using sanitation to remove pest habitat;
- using insect traps to monitor and control insect populations; and
- using physical barriers, such as row covers.

Examples of predatory beneficial insects include minute pirate bugs, big-eyed bugs, and to a lesser extent ladybugs (which tend to fly away), all of which eat a wide range of pests. Lacewings are also effective, but tend to fly away. Praying mantis tend to move more slowly and eat less heavily. Parasitoid wasps tend to be effective for their selected prey, but like all small insects can be less effective outdoors because the wind controls their movement. Predatory mites are effective for controlling other mites. Naturally derived insecticides allowed for use on organic farms use include Bacillus thuringiensis (a bacterial toxin), pyrethrum (a chrysanthemum extract), spinosad (a bacterial metabolite), neem (a tree extract) and rotenone (a legume root extract). Fewer than 10% of organic farmers use these pesticides regularly; one survey found that only 5.3% of vegetable growers in California use rotenone while 1.7% usepyrethrum.These pesticides are not always more safe or environmentally friendly than synthetic pesticides and can cause harm. The main criterion for organic pesticides is that they are naturally derived, and some naturally derived substances have been controversial. Controversial natural pesticides include rotenone, copper, nicotine sulfate, and pyrethrumsRotenone and pyrethrum are particularly controversial because they work by attacking the nervous system, like most conventional insecticides. Rotenone is extremely toxic to fish and can induce symptoms resembling Parkinson's disease in mammals. Although pyrethrum (natural pyrethrins) is more effective against insects when used with piperonyl butoxide (which retards degradation of the pyrethrins), organic standards generally do not permit use of the latter substance.

Naturally derived fungicides allowed for use on organic farms include the bacteria Bacillus subtilis and Bacillus pumilus; and the fungus Trichoderma harzianum. These are mainly effective for diseases affecting roots. Compost tea contains a mix of beneficial microbes, which may attack or out-compete certain plant pathogens, but variability among formulations and preparation methods may contribute to inconsistent results or even dangerous growth of toxic microbes in compost teas. Some naturally derived pesticides are not allowed for use on organic farms. These include nicotine sulfate, arsenic, and strychnine.

Synthetic pesticides allowed for use on organic farms include insecticidal soaps and horticultural oils for insect management; and Bordeaux mixture, copper hydroxide and sodium bicarbonate for managing fungi. Copper sulfate and Bordeaux mixture (copper sulfate plus lime), approved for organic use in various jurisdictions, can be more environmentally problematic than some synthetic fungicides dissallowed in organic farming Similar concerns apply to copper hydroxide. Repeated application of copper sulfate or copper hydroxide as a fungicide may eventually result in copper accumulation to toxic levels in soil, and admonitions to avoid excessive accumulations of copper in soil appear in various organic standards and elsewhere. Environmental concerns for several kinds of biota arise at average rates of use of such substances for some crops. In the European

Union, where replacement of copper-based fungicides in organic agriculture is a policy priority, research is seeking alternatives for organic production.

Genetic modification

Genetically modified crops, Genetically modified food, and Genetically modified food controversies A key characteristic of organic farming is the rejection of genetically engineered plants and animals. On 19 October 1998, participants at IFOAM's 12th Scientific Conference issued the Mar del Plata Declaration, where more than 600 delegates from over 60 countries voted unanimously to exclude the use of genetically modified organisms in food production and agriculture. Although opposition to the use of any transgenic technologies in organic farming is strong, agricultural researchers Luis Herrera-Estrella and Ariel Alvarez-Morales continue to advocate integration of transgenic technologies into organic farming as the optimal means to sustainable agriculture, particularly in the developing world. Organic farmer Raoul Adamchak and geneticist Pamela Ronald write that many agricultural applications of biotechnology are consistent with organic principles and have significantly advanced sustainable agriculture.

Although GMOs are excluded from organic farming, there is concern that the pollen from genetically modified crops is increasingly penetrating organic and heirloom seed stocks, making it difficult, if not impossible, to keep these genomes from entering the organic food supply. Differing regulations among countries limits the availability of GMOs to certain countries, as described in the article on regulation of the release of genetic modified organisms.

Tools

Organic farmers use a number of traditional farm tools to do farming. Due to the goals of sustainability in organic farming, organic farmers try to minimize their reliance on fossil fuels. In the developing world on small organic farms tools are normally constrained to hand tools and diesel powered water pumps.

Standards

Standards regulate production methods and in some cases final output for organic agriculture. Standards may be voluntary or legislated. As early as the 1970s private associations certified organic producers. In the 1980s, governments began to produce organic production guidelines. In the 1990s, a trend toward legislated standards began, most notably with the 1991 EU-Eco-regulation developed for European Union, which set standards for 12 countries, and a 1993 UK program. The EU's program was followed by a Japanese program in 2001, and in 2002 the

U.S. created the National Organic Program (NOP). As of 2007 over 60 countries regulate organic farming (IFOAM 2007:11). In 2005 IFOAM created the Principles of Organic Agriculture, an international guideline for certification criteria. Typically the agencies accredit certification groups rather than individual farms.

Composting

Using manure as a fertilizer risks contaminating food with animal gut bacteria, including pathogenic strains of E. coli that have caused fatal poisoning from eating organic food. To combat this risk, USDA organic standards require that manure must be sterilized through high temperature thermophilic composting. If raw animal manure is used, 120 days must pass before the crop is harvested if the final product comes into direct contact with the soil. For products that don't directly contact soil, 90 days must pass prior to harvest.

Economics

The economics of organic farming, a subfield of agricultural economics, encompasses the entire process and effects of organic farming in terms of human society, including social costs, opportunity costs, unintended consequences, information asymmetries, and economies of scale. Although the scope of economics is broad, agricultural economics tends to focus on maximizing yields and efficiency at the farm level. Economics takes an anthropocentric approach to the value of the natural world: biodiversity, for example, is considered beneficial only to the extent that it is valued by people and increases profits. Some entities such as the European Unionsubsidize organic farming, in large part because these countries want to account for the externalities of reduced water use, reduced water contamination, reduced soil erosion, reduced carbon emissions, increased biodiversity, and assorted other benefits that result from organic farming.

Traditional organic farming is labor and knowledge-intensive whereas conventional farming is capital-intensive, requiring more energy and manufactured inputs.

Geographic Producer Distribution

The markets for organic products are strongest in North America and Europe, which as of 2001 are estimated to have \$6 and \$8 billion respectively of the \$20 billion global market.As of 2007 Australasia has 39% of the total organic farmland, including Australia's 1,180,000 hectares (2,900,000 acres) but 97 percent of this land is sprawling rangeland (2007:35). US sales are 20x as much. Europe farms 23 percent of global organic farmland (6,900,000 ha (17,000,000 acres)), followed by

Latin America with 19 percent (5.8 million hectares - 14.3 million acres). Asia has 9.5 percent while North America has 7.2 percent. Africa has 3 percent.

Besides Australia, the countries with the most organic farmland are Argentina (3.1 million hectares - 7.7 million acres), China (2.3 million hectares - 5.7 million acres), and the United States (1.6 million hectares - 4 million acres). Much of Argentina's organic farmland is pasture, like that of Australia (2007:42). Spain, Germany, Brazil (the world's largest agricultural exporter), Uruguay, and the England follow the United States in the amount of organic land (2007:26).

In the European Union (EU25) 3.9% of the total utilized agricultural area was used for organic production in 2005. The countries with the highest proportion of organic land were Austria (11%) and Italy (8.4%), followed by the Czech Republic and Greece (both 7.2%). The lowest figures were shown for Malta (0.2%), Poland (0.6%) and Ireland (0.8%). In 2009, the proportion of organic land in the EU grew to 4.7%. The countries with highest share of agricultural land were Liechtenstein (26.9%), Austria (18.5%) and Sweden (12.6%). 16% of all farmers in Austria produced organically in 2010. By the same year the proportion of organic land increased to 20%.: In 2005 168,000 ha (415,000 ac) of land in Poland was under organic management. In 2012, 288,261 hectares (712,308 acres) were under organic production, and there were about 15,500 organic farmers; retail sales of organic products were EUR 80 million in 2011. As of 2012 organic exports were part of the government's economic development strategy.

After the collapse of the Soviet Union in 1991, agricultural inputs that had previously been purchased from Eastern bloc countries were no longer available in Cuba, and many Cuban farms converted to organic methods out of necessity. Consequently, organic agriculture is a mainstream practice in Cuba, while it remains an alternative practice in most other countries. Cuba's organic strategy includes development of genetically modified crops; specifically corn that is resistant to the palomilla moth.

Growth

In 2001, the global market value of certified organic products was estimated at USD $20 billion. By 2002, this was USD $23 billion and by 2015 more than USD $43 billion. By 2014, retail sales of organic products reached USD $80 billion worldwide. North America and Europe accounted for more than 90% of all organic product sales. In 2018 Australia accounted for 54% of the world's certified organic land with the country recording more than 35,000,000 verified organic hectares.

Organic agricultural land increased almost fourfold in 15 years, from 11 million hectares in 1999 to 43.7 million hectares in 2014.Between 2013 and 2014, organic agricultural land grew by 500,000 hectares worldwide, increasing in every region except Latin America. During this time period, Europe's organic farmland increased 260,000 hectares to 11.6 million total (+2.3%), Asia's increased 159,000 hectares to 3.6 million total (+4.7%), Africa's increased 54,000 hectares to 1.3 million total (+4.5%), and North America's increased 35,000 hectares to 3.1 million total (+1.1%).As of 2014, the country with the most organic land was Australia (17.2 million hectares), followed by Argentina (3.1 million hectares), and the United States (2.2 million hectares). Australia's organic land area has increased at a rate of 16.5% per annum for the past eighteen years. In 2013, the number of organic producers grew by almost 270,000, or more than 13%.By 2014, there were a reported 2.3 million organic producers in the world. Most of the total global increase took place in the Philippines, Peru, China, and Thailand. Overall, the majority of all organic producers are in India (650,000 in 2013), Uganda (190,552 in 2014), Mexico (169,703 in 2013) and the Philippines (165,974 in 2014).

Productivity

Studies comparing yields have had mixed results. These differences among findings can often be attributed to variations between study designs including differences in the crops studied and the methodology by which results were gathered. A 2012 meta-analysis found that productivity is typically lower for organic farming than conventional farming, but that the size of the difference depends on context and in some cases may be very small.While organic yields can be lower than conventional yields, another meta-analysis published in Sustainable Agriculture Research in 2015, concluded that certain organic on-farm practices could help narrow this gap. Timely weed management and the application of manure in conjunction with legume forages/cover crops were shown to have positive results in increasing organic corn and soybean productivity.

Another meta-analysis published in the journal Agricultural Systems in 2011 analyzed 362 datasets and found that organic yields were on average 80% of conventional yields. The author's found that there are relative differences in this yield gap based on crop type with crops like soybeans and rice scoring higher than the 80% average and crops like wheat and potato scoring lower. Across global regions, Asia and Central Europe were found to have relatively higher yields and Northern Europe relatively lower than the average.

A 2007 study compiling research from 293 different comparisons into a single study to assess the overall efficiency of the two agricultural systems has concluded that "organic methods could produce enough food on a global per capita basis to sustain the current human population, and potentially an even larger population, without increasing the agricultural land base." The researchers also found that while in developed countries, organic systems on average produce 92% of the yield produced by conventional agriculture, organic systems produce 80% more than conventional farms in developing countries, because the materials needed for organic farming are more accessible than synthetic farming materials to farmers in some poor countries. This study's methodology and results were contested by D.J. Connor of The University of Melbourne, in a short communication published in Field Crops Research. Connor writes that errors in Badgley et al. result in "major overestimation of the productivity of OA".

Long term studies

A study published in 2005 compared conventional cropping, organic animal-based cropping, and organic legume-based cropping on a test farm at the Rodale Institute over 22 years.The study found that "the crop yields for corn and soybeans were similar in the organic animal, organic legume, and conventional farming systems". It also found that "significantly less fossil energy was expended to produce corn in the Rodale Institute's organic animal and organic legume systems than in the conventional production system. There was little difference in energy input between the different treatments for producing soybeans. In the organic systems, synthetic fertilizers and pesticides were generally not used". As of 2013 the Rodale study was ongoing and a thirty-year anniversary report was published by Rodale in 2012. A long-term field study comparing organic/conventional agriculture carried out over 21 years in Switzerland concluded that "Crop yields of the organic systems averaged over 21 experimental years at 80% of the conventional ones. The fertilizer input, however, was 34 – 51% lower, indicating an efficient production. The organic farming systems used 20 – 56% less energy to produce a crop unit and per land area this difference was 36 – 53%. In spite of the considerably lower pesticide input the quality of organic products was hardly discernible from conventional analytically and even came off better in food preference trials and picture creating methods"

EXTERNALITIES

Agriculture imposes negative externalities (uncompensated costs) upon society through public land and other public resource use, biodiversity loss, erosion, pesticides, nutrient runoff, subsidized water usage, subsidy payments and assorted other problems. Positive externalities include self-reliance, entrepreneurship, respect for nature, and air quality. Organic methods reduce some of these costs.

In 2000 uncompensated costs for 1996 reached 2,343 million British pounds or £208 per ha (£84.20/ac).A study of practices in the US published in 2005 concluded that cropland costs the economy approximately 5 to 16 billion dollars ($30–96/ha – $12–39/ac), while livestock production costs 714 million dollars.Both studies recommended reducing externalities. The 2000 review included reported pesticide poisonings but did not include speculative chronic health effects of pesticides, and the 2004 review relied on a 1992 estimate of the total impact of pesticides.

It has been proposed that organic agriculture can reduce the level of some negative externalities from (conventional) agriculture. Whether the benefits are private or public depends upon the division of property rights.

Several surveys and studies have attempted to examine and compare conventional and organic systems of farming and have found that organic techniques, while not without harm, are less damaging than conventional ones because they reduce levels of biodiversity less than conventional systems do and use less energy and produce less waste when calculated per unit area.

Disadvantages

A 2003 to 2005 investigation by the Cranfield University for the Department for Environment, Food and Rural Affairs in the UK found that it is difficult to compare the Global warming potential, acidification and eutrophication emissions but "Organic production often results in increased burdens, from factors such as N leaching and N2O emissions", even though primary energy use was less for most organic products. N2O is always the largest global warming potential contributor except in tomatoes. However, "organic tomatoes always incur more burdens (except pesticide use)". Some emissions were lower "per area", but organic farming always required 65 to 200% more field area than non-organic farming. The numbers were highest for bread wheat (200+ % more) and potatoes (160% more).

Environmental Impact and Emissions

Researchers at Oxford University analyzed 71 peer-reviewed studies and observed that organic products are sometimes worse for the environment. Organic milk, cereals, and pork generated higher greenhouse gas emissions per product than conventional ones but organic beef and olives had lower emissions in most studies.Usually organic products required less energy, but more land. Per unit of product, organic produce generates higher nitrogen leaching, nitrous oxide emissions, ammonia emissions, eutrophication, and acidification potential than conventionally grown produce.Other differences were not significant. The researchers concluded that public debate should consider various manners of employing conventional or organic farming, and not merely debate conventional farming as opposed to organic farming. They also sought to find specific solutions to specific circumstances.

Proponents of organic farming have claimed that organic agriculture emphasizes closed nutrient cycles, biodiversity, and effective soil management providing the capacity to mitigate and even reverse the effects of climate change and that organic agriculture can decrease fossil fuel emissions. "The carbon sequestration efficiency of organic systems in temperate climates is almost double (575–700 kg carbon per ha per year – 510–625 lb/ac/an) that of conventional treatment of soils, mainly owing to the use of grass clovers for feed and of cover crops in organic rotations."

Critics of organic farming methods believe that the increased land needed to farm organic food could potentially destroy the rainforests and wipe out many ecosystems.

Nutrient Ieaching

According to a 2012 meta-analysis of 71 studies, nitrogen leaching, nitrous oxide emissions, ammonia emissions, eutrophication potential and acidification potential were higher for organic products, although in one study "nitrate leaching was 4.4–5.6 times higher in conventional plots than organic plots". Excess nutrients in lakes, rivers, and groundwater can cause algal blooms, eutrophication, and subsequent dead zones. In addition, nitrates are harmful to aquatic organisms by themselves.

Land use

The Oxford meta-analysis of 71 studies found that organic farming requires 84% more land for an equivalent amount of harvest, mainly due to lack of nutrients but sometimes due to weeds, diseases or pests, lower yielding animals and land required for fertility building crops. While organic farming does not necessarily save land for wildlife habitats and forestry in all cases, the most modern breakthroughs in organic are addressing these issues with success. Professor Wolfgang Branscheid says that organic animal production is not good for the environment, because organic chicken requires twice as much land as "conventional" chicken and organic pork a quarter more. According to a calculation by Hudson Institute, organic beef requires three times as much land. On the other hand, certain organic methods of animal husbandry have been shown to restore desertified, marginal, and/or otherwise unavailable land to agricultural productivity and wildlife. Or by getting both forage and cash crop production from the same fields simultaneously, reduce net land use. In England organic farming yields 55% of normal yields. In other regions of the world, organic methods have started producing record yields.

Pesticides

In organic farming synthetic pesticides are generally prohibited. A chemical is said to be synthetic if it does not already exist in the natural world. But the organic label goes further and usually prohibit compounds that exist in nature if they are produced by chemical synthesis. So the prohibition is also about the method of production and not only the nature of the compound. A non-exhaustive list of organic approved pesticides with their median lethal doses:

- Copper(II) sulfate is used as a fungicide and is also used in conventional agriculture (LD50 300 mg/kg). Conventional agriculture has the option to use the less toxic Mancozeb (LD50 4,500 to 11,200 mg/kg)
- Boric acid is used as stomach poison that target insects (LD50: 2660 mg/kg).
- Pyrethrin comes from chemicals extracted from flowers of the genus Pyrethrum (LD50 of 370 mg/kg). Its potent toxicity is used to control insects.
- Lime sulphur (aka calcium polysulfide) and sulfur are considered to be allowed, synthetic materials (LD50: 820 mg/kg)
- Rotenone is a powerful insecticide that was used to control insects (LD50: 132 mg/kg). Despite the high toxicity of Rotenone to aquatic life and some links to Parkinson disease the compound is still allowed in organic farming as it is a naturally occurring compound.
- Bromomethane is a gas that is still used in the nurseries of Strawberry organic farming
- Azadirachtin is a wide spectrum very potent insecticide. Almost non toxic to mammals (LD50 in rats is > 3,540 mg/kg) but affects beneficial insects.

Food quality and safety

While there may be some differences in the amounts of nutrients and anti-nutrients when organically produced food and conventionally produced food are compared, the variable nature of food production and handling makes it difficult to generalize results, and there is insufficient evidence to make claims that organic food is safer or healthier than conventional food. Claims that organic food tastes better are not supported by evidence.

Soil conservation

Supporters claim that organically managed soil has a higher quality and higher water retention. This may help increase yields for organic farms in drought years. Organic farming can build up soil organic matter better than conventional no-till farming, which suggests long-term yield benefits from organic farming. An 18-year study of organic methods on nutrient-depleted soil concluded that conventional methods were superior for soil fertility and yield for nutrient-depleted soils in cold-temperate climates, arguing that much of the benefit from organic farming derives from imported materials that could not be regarded as self-sustaining.

In Dirt: The Erosion of Civilizations, geomorphologist David Montgomery outlines a coming crisis from soil erosion. Agriculture relies on roughly one meter of topsoil, and that is being depleted ten times faster than it is being replaced.No-till farming, which some claim depends upon pesticides, is one way to minimize erosion. However, a 2007 study by the USDA's Agricultural Research Service has found that manure applications in tilled organic farming are better at building up the soil than no-till.

Biodiversity

The conservation of natural resources and biodiversity is a core principle of organic production. Three broad management practices (prohibition/reduced use of chemical pesticides and inorganic fertilizers; sympathetic management of non-cropped habitats; and preservation of mixed farming) that are largely intrinsic (but not exclusive) to organic farming are particularly beneficial for farmland wildlife. Using practices that attract or introduce beneficial insects, provide habitat for birds and mammals, and provide conditions that increase soil biotic diversity serve to supply vital ecological services to organic production systems. Advantages to certified organic operations that implement these types of production practices include: 1) decreased dependence on outside fertility inputs; 2) reduced pest management costs; 3) more reliable sources of clean water; and 4) better pollination.

Nearly all non-crop, naturally occurring species observed in comparative farm land practice studies show a preference for organic farming both by abundance and diversity. An average of 30% more species inhabit organic farms. Birds, butterflies, soil microbes, beetles, earthworms, spiders, vegetation, and mammals are particularly affected. Lack of herbicides and pesticides improve biodiversity fitness and population density.Many weed species attract beneficial insects that improve soil qualities and forage on weed pests. Soil-bound organisms often benefit because of increased bacteria populations due to natural fertilizer such as manure, while experiencing reduced intake of herbicides and pesticides. Increased biodiversity, especially from beneficial soil microbes and mycorrhizae have been proposed as an explanation for the high yields experienced by some organic plots, especially in light of the differences seen in a 21-year comparison of organic and control fields.

Biodiversity from organic farming provides capital to humans. Species found in organic farms enhance sustainability by reducing human input (e.g., fertilizers, pesticides). The USDA's Agricultural Marketing Service (AMS) published a Federal Register notice on 15 January 2016, announcing the National Organic Program (NOP) final guidance on Natural Resources and Biodiversity Conservation for Certified Organic Operations. Given the broad scope of natural resources which includes soil, water, wetland, woodland and wildlife, the guidance provides examples of practices that support the underlying conservation principles and demonstrate compliance with USDA organic regulations § 205.200. The final guidance provides organic certifiers and farms with examples of production practices that support conservation principles and comply with the USDA organic regulations, which require operations to maintain or improve natural resources. The final guidance also clarifies the role of certified operations (to submit an OSP to a certifier), certifiers (ensure that the OSP describes or lists practices that explain the operator's monitoring plan and practices to support natural resources and biodiversity conservation), and inspectors (onsite inspection) in the implementation and verification of these production practices.

A wide range of organisms benefit from organic farming, but it is unclear whether organic methods confer greater benefits than conventional integrated agri-environmental programs. Organic farming is often presented as a more biodiversity-friendly practice, but the generality of the beneficial effects of organic farming is debated as the effects appear often species- and context-dependent, and current research has highlighted the need to quantify the relative effects of local- and landscape-scale management on farmland biodiversity. There are four key issues when comparing the impacts on biodiversity of organic and conventional farming: (1) It remains unclear whether a holistic whole-farm approach (i.e. organic) provides greater benefits to biodiversity than carefully targeted prescriptions applied to relatively small areas of cropped and/or non-cropped habitats within conventional agriculture (i.e. agri-environment schemes); (2) Many comparative studies encounter methodological problems, limiting their ability to draw quantitative conclusions; (3) Our knowledge of the impacts of organic farming in pastoral and upland agriculture is limited; (4) There remains a pressing need for longitudinal, system-level studies in order to address these issues and to fill in the gaps in our knowledge of the impacts of organic farming, before a full appraisal of its potential role in biodiversity conservation in agroecosystems can be made.

5

POST HARVEST MANAGEMENT OF HORTICULTURAL CROPS

This chapter does not attempt to cover in depth the technical details of post-harvest handling of horticultural crops. FAO and others have covered this in many other publications. The main emphasis here is on the commercial implications of different harvest and post-harvest practices. The key issues are:

- Maintaining quality;
- Maximizing shelf life;
- Supplying produce when and as the market demands it.

The main way that you can help improve post-harvest handling is by training farmers, both through practical training and by presentations. In the following pages the main ways in which post-harvest techniques can increase farmers' profits are set out.

HARVESTING

The timing, techniques and conditions of harvesting can significantly affect prices. Harvesting and prices. Harvesting early in the season can be carried out to take advantage of opportunities for high prices, e.g. cabbage harvested as spring greens, young carrots sold in bunches, green plums and new potatoes. Taking advantage of these short-term market opportunities requires close links with the market. Harvesting and crop maturity. The shelf life of the crop and its suitability for long-term storage is affected by the maturity of the crop at harvest. The optimum harvesting stage for most crops depends not only on the climate and distance to the market but also on the variety and the growing conditions. When distant markets are being investigated, experiments should be carried out to find the best maturity to harvest fruits. Send samples at different degrees of ripeness and assess which gives the best results. It may be necessary to call in expert assistance to identify whether long-term crop storage could significantly improve farmer incomes.

Harvesting and quality. Growers often do not understand the effect of their harvesting and handling on the quality of the produce when it reaches the market. Once a fruit is plucked from a plant, or a root or leaf vegetable is harvested, it is cut off from its source of food and, particularly, water. The effects of poor treatment normally show themselves some days later, when the produce is being presented for sale or is in storage. This can often result in disputes, because farmers have sent to market what they consider to be good quality products but by the time these arrive at the market the trader sees produce that has deteriorated badly. An example of this comes from Tonga, in the South Pacific, which used to export capsicum (green pepper) to New Zealand. When the fruit left Tonga they appeared to be in excellent condition but after four days in a ship they arrived in very poor condition. The problem was traced to the use of dirty knives for harvesting.

Harvesting and Handling

Quality in produce can be defined as the composite of characteristics that differentiates individual items within a commodity and have a significant influence in determining the level of acceptance by the consumer. High quality vegetables are one of today's wonders with regard to the U.S. food supply. Locally produced vegetables, readily available seasonally have resulted in consumers demanding the same freshness and quality on a year round basis. Unfortunately, many Texas vegetable growers believe that once a high quality product is produced, their problems are over. In reality, their troubles could be just beginning. It has been estimated that more than 40% of perishable commodities are lost after production.

Fresh vegetables are extremely perishable and have relatively short shelf lives. They are living, respiring tissues that start senescing immediately at harvest. Freshly harvested vegetables are mostly comprised of water with most having 90 to 95% moisture content. Water loss after harvest is one of the most serious postharvest conditions. Consequently, special effort is required to reduce the effects of these naturally occurring processes if quality harvested in the field will be the same at the consumer level.

Maintaining Quality

Special skills are required for proper harvesting, handling, grading and packaging of vegetables in order to insure optimum produce quality at the marketplace. It makes little difference what the quality is at harvest if it is reduced by poor handling, packaging or storage conditions. Price received for produce is determined by quality at the marketplace. Variables consumers perceive as a reflection of produce quality are ranked in order of preference as follows: crispness and freshness, taste, appearance and condition, nutritive value, and price. Studies have shown that two factors normally enter into consumers purchase decisions: competition between like items on the display shelf, and, the acceptability of the item in reference to his or her standard for that item in reference to the above variables. Consequently, producer who are able to produce and package their produce in such a way to enhance these variables are the most successful in the market place.

Because of the perishable nature of vegetables, harvesting and handling speed is of utmost importance as soon as harvest maturity has occurred. Every producer should have products reach the end consumer as quickly as possible. Unfortunately, they have no control once the produce leaves their farm or packing sheds. However, maximum speed and efficiency in handling produce on the farm or at the shed will help maintain quality. Consequently, a grower must be prepared to operate in advance of the actual harvest operation. Preharvest preparation should include lining up sufficient labor, supplies (containers and packaging items), cleaning the grading/ packing shed, and determining if all equipment is operable. Once the produce reaches harvest maturity, delays for any reason can result in major quality and crop losses. In addition, the nutritive content of produce is not static either. Biosynthetic and degradation reactions will even continue to occur during handling and storage.

Techniques to extend postharvest shelf life:

- Reducing respiration by lowering temperature
- Slow respiration by maintaining optimal gaseous environment
- Slow water loss by maintaining optimal relative humidity
- Variety selection

Postharvest handling includes all steps involved in moving a commodity from the producer to the consumer including harvesting, handling, cooling, curing, ripening, packing, packaging, storing, shipping, wholesaling, retailing, and any other procedure that the product is subjected to. Because vegetables can change hands so many times in the Postharvest sector, a high level of management is necessary to ensure that quality is maintained. Each time someone fails to be conscientious in carrying out his or her assigned responsibility, quality is irreversibly sacrificed. Maintaining produce quality from the farm to the buyer is a major prerequisite of successful marketing. The initial step required to insure successful marketing is to harvest the crop at the optimum stage of maturity. Full red, vine-ripened tomatoes may be ideal to meet the needs of a roadside stand, but totally wrong if the fruit is destined for long distance shipment. Factors such as size, color, content of sugar, starch, acid, juice or oil, firmness, tenderness, heat unit accumulation, days from bloom, and specific gravity can be used to schedule harvest. Vegetable producers should gather as much information as possible on maturity indices for their particular commodities. The result of harvesting at an inappropriate stage of development can be a reduction in quality and yield.

REDUCING DAMAGE

Because of stressful conditions during the harvest season in most parts of Texas, proper timing of the harvest period is essential. For vegetables that lose quality rapidly due to high temperatures, such as sweet corn and green beans, early morning harvest prior to the buildup of field heat may be in order. However, with some vegetables which are prone to breakage, such as asparagus, late afternoon harvest when wilting generally occurs may be more beneficial.

Most postharvest problems are management problems, and therefore, are people problems. The objective of this chapter is to identify the most common problems in postharvest management and to emphasize what those involved in handling systems can do to ensure that consumers receive products of the highest quality. With all vegetables, care should be taken to prevent injury due to harvesting and handling errors. A crucial time to be aware of this is during the harvesting operation. The hands of the harvest worker are the most important hands that ever touch the product. The truly skilled worker does not inflict injury to the product. Broken skin and bruises reduce eye appeal and provide a ready access to decay organisms and enhance physiological breakdown. Although speed is an important consideration, excessive, unsupervised speed may result in a greater incidence of injury and quality losses. Therefore, time should be spent to properly train and monitor the performance of all personnel to insure maximum efficiency without sacrificing quality. Equally important is the need for periodic inspection and repair of all harvest containers, bulk bins and grading equipment to insure that these items are not causing injury to the produce. Bruise damage will cause respiration rates and ethylene production to increase dramatically. This shortens the shelf life. There are several management practices that can reduce or eliminate harvest injury.

- Remove protruding nails or staples and smooth the rough edges on field containers.
- Harvest workers should not have long, sharp fingernails.
- Use care in dumping products from one container to another. This is one of the most common trouble spots.
- Use padding on all impact areas when possible.
- Clean sand and all debris out of all containers.
- Don't overfill containers! Severe damage can result when stacked.
- Consider the time of day of harvest. Many products are more turgid in the early morning and bruise more easily.

Transport from field to packinghouse can be a source of injury. Roads should be maintained in good condition. Drivers should exercise care and remember that they are transporting living material. The springs and shock absorbers on trucks and trailers must be properly maintained.

Dumping or unloading at the packing shed is also a trouble spot. Dry dumping is an option for certain produce items. If a wet dump is necessary, appropriate flow control out of the dumping area is needed to minimize bruising. The packing line itself should have as few drops and shears as possible. Shears that are essential should be designed properly.

Precooling

Rapid cooling as soon as possible after harvest is essential to the maintenance of optimum quality. The first consideration at harvest is removal of the produce from direct sunlight, and secondly, to precool as quickly as possible.

PRECOOLING METHODS

Room Cooling

Exposure of produce to cold air in an enclosed space is the simplest and most common cooling method. Cold air normally is discharged horizontally near the ceiling so as to enable it to return through produce stacked on the floor. The precooking room may also be the storage room. This results in less handling of the product. Heat is removed slowly so the peak demand for refrigeration is less than with more rapid cooling systems. Since cooling is slow, shipments may be delayed, or in some cases the product may be shipped without adequate precooking. Certain commodities, such as snap beans, may deteriorate before cooling is accomplished. These problems may be minimized by ensuring that containers are stacked to facilitate good air circulation. Fans must be powerful enough to move the air at a velocity of 2 to 4 miles per hour among the containers, which should be vented adequately.

Forced-Air Cooling

Forced-air cooling or pressure cooling is accomplished through the use of fans and strategically placed barriers so that cold air is forced to pass through the containers of produce. This method usually takes from 1/4th to 1/10th the time required to cool produce by passive room cooling, but takes two or three times longer than hydro or vacuum cooling. Room coolers are relatively easy to adapt to forced-air. However, the refrigeration capacity of the room may need to be increased to compensate for the rapid removal of heat from the produce. Commodities should not be left on the cooler longer than the time required to reach about 1/8th of their initial temperature because water loss from the product is increased. The importance of humidity control is discussed in a later section. Cooling schedules should be calculated to avoid this problem.

Icing

The use of ice for cooling, either by package icing or by bulk application to the top of a load, is one of the oldest cooling methods. The use of slurries and mechanization of the process have made this a popular method for some commodities like broccoli. Cartons must be able to withstand free water. Once iced, they should be placed in a refrigerated storage unit or transit vehicle. As the ice melts, high relative humidity is present surrounding the produce. A rule of thumb for the cooling capacity of ice is that, to cool a produce from 95F to 35F, the weight of ice needed is equivalent to 38% of the produce weight. Most packed cartons have enough free space to include this amount of ice. It is essential that only high quality ice made from potable water be used.

Hydrocooling

Hydrocooling is one of the most efficient of all methods for precooking. Produce is drenched with cold water, either on a moving conveyor or in a stationary setting. In some cases, commodities may be forced through a tank of cold water. Hydrocooling is an excellent method for bulky items such as sweet corn, peaches, or cantaloupes. Good water sanitation practices must be observed and once cooled, the produce should be kept cold. The cold water must come in direct contact with the product, so it is essential the containers be designed and filled in such a way that the water does not simply channel through without making contact. Studies have shown that some commercial hydrocoolers are not adequately insulated and as much as half of the energy for refrigeration is wasted.

Vacuum Cooling

Commodities may be enclosed in a sealed container from which air and water vapor are rapidly pumped out. As the air pressure is reduced, the boiling point of water is lowered, so the product is cooled by surface water evaporation. Vacuum cooling works best with products that have a high surface to volume ratio, such as lettuce or leafy greens. The method is effective on produce that is already packaged providing there is a means for water vapor to escape. Moisture loss from the commodity is generally within the range of 1.5 to 5.0%. Generally, about 1% of the weight is lost for each 10oF the product is cooled. This can be reduced by wetting the product before cooling. Vacuum chambers vary in size from very large, about rail car size, to smaller portable units that may be taken to the field.

Evaporative Cooling

The use of wet pads and fans to cool the air in greenhouses is a common practice. This can be adapted to the cooling of horticultural commodities; however, the benefits are few compared to other methods of cooling. With this method air temperature can be reduced to near the wet bulb temperature, but not below it. Therefore, most items cannot be cooled adequately by this method. It does offer the benefit of surrounding the produce with air of very high relative humidity, which helps to reduce shrinkage. Roadside markets, or other outlets in which long term storage is not needed, may be the most practical application for an evaporative cooler.

The cooling methods described here represent only a few of the available temperature management procedures. Once a product is cooled to the lowest safe temperature, it should be kept cold. Transit vehicles, such as trucks, rail cars, etc. should be precooled before loading. Their refrigeration units are not designed to remove field heat, only to maintain the low temperature during transit. Forklift operators and other handlers should be aware that undue exposure to warm air will result in loss of quality. For some commodities, a few hours at warm temperatures can result in some loss of the product.

The process by which harvested vegetables sustain life from its stored food reserves is termed respiration. An understanding of respiration is useful in estimating the need for refrigeration since a portion of the energy generated by respiration is given off as heat. The amount of heat given off is a function of both the respiration rate of the commodity and the temperature at which it is stored. In general, the respiration rate doubles with each 10C (18F) rise in temperature, and as the respiration rate increases, the shelf life decreases. It should be obvious that the respiration rate and the heat generated by respiration can be minimized by appropriate temperature management throughout all steps in the postharvest handling scheme. In fact, temperature is the single most important factor to control in the postharvest environment. To ignore this is to lose money.

Food Safety

Environmental activists and the media have raised serious concerns over the safety of our food supply. Their concerns have been focused on pesticide use on and residues in our fruits and vegetables. However, a more eminent danger lies in the potential for borne illness due to microbial contamination of fresh produce. These organisms can readily contaminate fresh produce during any number of operations associated with the production, handling, packaging, processing, distribution, and preparation of produce for consumption. As a result, significantly higher incidences of illnesses and deaths have been found to be caused by microbial contamination than from pesticide poisoning. The danger of contamination of produce by microbial human pathogens is increased with the use of untreated manures or improperly handled compost on crops grown in, on or near the soil surface. Such products area potent source of the potentially deadly bacteria Escherichia coli 0157:h7, Salmonella and Cryptosporidium. Consequently, another important facet of market quality demands is clean fruit. Most vegetables should be washed after harvest to remove dirt, pesticide residue and to freshen up wilted items. With the proper design of a produce shed, the washing and cooling processes can be combined into one continuous operation. Proper design and maintenance of packing sheds can go a long way in reducing the incidence of contaminated produce being shipped to supermarkets. The perceived seriousness of the problem has caused many of the nation's largest supermarket chains to demand that growers or grower/shippers to have an independent third party audit of their operation and facilities to insure their products are grown, packed, and shipped using good agricultural practices that reduce the incidences of microbial contamination.

The Food and Drug Administration in the interest of protecting our nation's food supply has developed a set of guidelines for minimizing microbial food safety hazards for fresh fruits and vegetables.

Food Safety Guidelines

Agricultural Water: Water quality should be adequate for its intended use. Where water quality is unknown or cannot be controlled, growers should use good management practices which minimize the risk of contamination such as protecting surface waters, wells and pump areas from uncontrolled livestock or wildlife access to limit the extent of fecal contamination. Grass sod waterways, diversion berms, runoff control structures, and vegetative buffer areas may prevent polluted runoff water from contaminating agricultural water sources and produce crops. Where water quality is unknown, try to avoid direct contact of the water to editable crops. Test water source(s) frequently for microbial contamination.

Manure and Municipal Biosolids: Use only properly treated manure and biosolids to fertilize vegetables. Situate, if practical, vegetable production fields away from other potential sources of contamination such as nearby composting or manure storage areas, livestock or poultry operations, nearby municipal wastewater or biosolids storage, treatment or disposal areas, and high concentrations of wildlife in the growing and harvesting and packing environment. Worker Health and Hygiene: Establish hygiene practices for worker appropriate for the agricultural environment (field, packing shed and transportation operation) and provide worker training on the hygiene practices required. Contact of fresh produce with fecal matter is the major cause of food borne illnesses. Most of which is a result of the lack of proper worker hygiene practices. Also infectious diseases, accompanied by diarrhea or open lesions, that include boils, sores, or infected wounds are a source of disease causing microorganisms. Sick or injured workers with these symptoms should be requested to remain at home or be assigned to duties in which they do not come in contact with fresh produce. Sanitary Facilities: Producers should describe the appropriate number of toilets to the number of workers, proper hand washing facilities, maximum worker to restroom distance, how often such a facility should be cleaned, and how the waste is to be disposed to avoid contamination of worker and produce. Field Stations: Attempts should be made to prevent cross-contamination of fresh produce during preharvest and harvest activities that may result from contact with soils, fertilizers, water, workers, and harvesting equipment. Clean harvest containers and storage facilities prior to use, discard broken or damage bins, field containers and bulk load equipment periodically, remove as much dirt as possible in the field, and maintain harvest equipment.

Packing Sheds: Maintain buildings, fixtures, and other physical facilities, and their grounds. Practice good sanitation within the packing shed. Clean pallets, rest rooms, containers, grading and packing lines daily. Develop a vigorous pest control program. Sanitation within the Shed: Problems of rough handling and poor temperature management are compounded when poor sanitation practices exist in the postharvest environment. Packing sheds and storerooms should always be clean and neat. Products left on the floor under machinery will rot and contaminate the air with spores of decay causing organisms that may then infect other commodities. The packing line itself should be left free of produce each day and cleaned regularly. Bulk bins, storage buckets, and other containers should be cleaned and disinfected regularly. The risk of decay always increases with the exposure of commodities to water. Dump tanks, which are used to minimize bruising during dumping of fruits and vegetables, can be a particular problem if not managed properly.

The following general management practices should be observed:

- Start with clean water each day.
- Chlorinate to approximately 100 ppm free chlorine. This may be done with chlorine gas or with either liquid or granular hypochlorite.
- Adjust the pH to the range of 6.0 to 7.0. Chlorine gas has little effect on pH but hypochlorites raise pH, so some adjustment is usually necessary. Use caution of strong acid or base is used to make this adjustment.
- For certain commodities, especially tomatoes, the water should be heated about 10F warmer than the fruit.
- Minimize exposure time in the tank. Never load the tank with produce unless the packing line is working.
- Clean out the tank thoroughly at the end of each day.
- Double check all of the above frequently.
- Transport vehicles should always be cleaned and sanitized before loading.

The same general rules for managing dump tanks also apply to hydro coolers. Even though microbial growth is slower at low temperature, refrigeration does not kill microorganisms, so hydro coolers can spread decay in the same manner as a dump tank. The condition of rinse water, e.g. chlorination pH, etc., is not as critical as dump tank or hydrocooler water because the fruit is not immersed. However, any time commodities are in contact in water, it is wise to practice good sanitation.

Adjustment of water pH has been a major shortcoming of packinghouse managers in South Carolina, even though pH has a critical effect on the efficiency of chlorine to disinfect. Equipment now is available for automatic control of pH and chlorine content. Many vegetables are not treated with fungicides or other means of chemical decay control. Since we have no way to cure a decay problem, we must prevent it from occurring. Good sanitation, in addition to appropriate temperature management and handling practices, may be the packers' only resource for decay control.

Packing

Appearance plays a major role in vegetable sales success. Therefore, a grower should pay special attention to maturity, size, color, shape, and freedom from blemishes and dirt when grading and sorting produce. Each package or display bin should contain fruit having similar qualities. If vegetables are to be shipped, container size is an important consideration. A good rule of thumb is, the further a grower is from the market, the greater the marketing cost, and the greater the requirement for careful grading and packaging. Table 43 in the Appendix, lists commonly used containers and weights for fresh vegetables. A good package should protect the product during storage and transit, facilitate temperature management, protect from water loss, be compatible with any special treatments required by the product, and be compatible to existing handling systems. Standardization of containers has been a major challenge for Post harvest technologies and little progress has been made, in practice, to reduce the variety of packages in use.

One of the newest developments in packaging has been the use of materials that allow for modification of the atmosphere within the package, thus the term modified-atmosphere (MA) packaging. This involves the use of plastic films that allow depletion of oxygen and accumulation of carbon dioxide within the container, which will increase the shelf life of some commodities. However, some products develop off-odors or have increased bacterial growth under MA, and much research is ongoing to define which products respond best to MA. It should be noted that MA will not substitute for good temperature management. The high cost of MA may be wasted if other aspects of Post harvest management are ignored.

Storage

Proper storage and/or transit conditions should be used if vegetables are to be held prior to sale or if they are to be transported over a great distance.

One of the major problems encountered during storage of certain vegetables is chilling injury.

Another important consideration in order to maintain optimum storage conditions is relative humidity. Relative humidity is defined as the amount of water present in air relative to the maximum amount that the air can hold at that particular temperature. It is usually expressed as a percentage. Small fluctuations in temperature can cause wide fluctuations in relative humidity. This critical component of the Post harvest environment is often overlooked. Products stored at less than optimum relative humidity will suffer excessive water loss and begin to shrivel. Many vegetables are unacceptable for marketing if weight loss reaches 5% because of their undesirable appearance and undesirable textural changes that may accompany water loss. Leafy vegetables are among the less tolerant crops to dehydration.

The use of large refrigeration coils for heat exchange allows them to be operated with fewer differentials between coil and air temperature, thereby reducing the amount of moisture removed from the room by condensation on the coil. However, under almost all circumstances, water vapor needs to be added back to the air. There are a number of humidifying systems available for this purpose. In the absence of a humidifying system, water can be applied to the floor and walls to temporarily increase relative humidity. For long term storage, automated systems are essential. Managers of packing houses, storage facilities, etc. should have at their disposal some means of measuring relative humidity.

RECOMMENDATIONS FOR SELECTING VEGETABLES

Bean

Beans are subject to chill injury and russeting if held at temperatures below 40°F. Chill damage may begin to show within three days. If held at 32°F they will russet and become pitted and lose moisture rapidly. Optimum temperatures range between 45 and 50°F. Beans stored too long or at too high a temperature are subject to various soft rots. Washing beans before refrigerating will help to retain moisture content. Snap beans require air circulation. Containers should be stacked to allow maximum air circulation.

Broccoli

The recommended temperature for storing broccoli is 32°F. If in good condition and stored with adequate air circulation, broccoli will last 10 to 14 days. Storage beyond this time can cause leaves to discolor, buds to drop off, turn yellow and tissue to deteriorate. Age can cause stem hardening and/or decay. To avoid dehydration, sprinkle crushed ice or water on the broccoli as soon after harvest as possible. Broccoli should be iced when transporting to markets.

Cabbage

Cabbage should be stored at 32°F. Cabbage loses moisture easily if held at room temperature. Cut heads do not hold up well in storage. If held at 32°F, cabbage will last 10 to 14 days.

Cantaloupe

These melons are highly perishable and the shortest possible storage time is recommended. Temperatures should be held between 40 and 50°F. The melon must mature on the vine before it is harvested. It does not increase in sugar content after picking, although it will soften and change types of sugars. Only a mature melon will provide the sweetness, texture, flavor and juiciness that characterize cantaloupes at their finest. A mature cantaloupe is well netted or webbed with a smooth rounded, depressed scar at the tip end. Cantaloupes for shipping are usually harvested while firm and prior to reaching full maturity in order to avoid damage during transit. A few days at room temperature will soften the fruit. Cantaloupes should be carefully handled to avoid damage. Bruising, scuffing of the netting, puncturing or cracking can cause more rapid deterioration through moisture loss or decay. Under optimum conditions a cantaloupe will deteriorate from field fresh to fair within 12 to 16 days. The operator can expect a 5 to 7 day shelf life. Because cantaloupes generate a significant amount of ethylene gas mixed load shipment with ethylene sensitive produce should be avoided.

Carrot

The recommended temperature for storing carrots is 32°F. Carrots will wilt and dry if not kept sufficiently moist. When properly handled, carrots can be stored for several months. Fresh market carrots generally are harvested before reaching full maturity. Avoid storing with ethylene producing commodities such as cantaloupes, tomatoes, and most fruits.

Cauliflower

Cauliflower should not be held for any length of time. However, if in good condition, with temperatures held at 32°F, it will keep for several weeks. Cauliflower should be stored with the head down to prevent accumulation of moisture on the curds. A good quality head of cauliflower will be white or creamy white in color. Curds should be clean, firm and compact with jacket leaves fresh and green. Small leaves extending through the curd do not influence the eating quality. Ethylene gas will accelerate decay and brown spotting. Do not store or ship with ethylene producing produce.

Celery

If held at 32°F, celery will keep well in storage. Celery should not be sprinkled if it is packaged, which encourages decay. Celery should be brittle enough to break easily with fresh, crisp and clean stalks. The inside stem should be smooth. If it feels rough or puffy, the celery is likely to be pithy. Leaflets should be fresh or only slightly wilted. Light green stalks with glossy surfaces offer the best taste.

Cucumber

Cucumbers will keep well for 10 to 14 days when temperatures are maintained between 45 and 50°F. Temperatures below 45F for more than two days will cause chill damage. Chill injured cucumbers will develop water-soaked spots, pitting or tissue collapse. Extensive decay will develop when removed from low temperatures. Temperatures above 50F will cause the product to ripen rapidly and take on a yellow coloring after about 10 days. Ripening will be accelerated if cucumbers are stored with ethylene producing commodities. Adequate humidity is also important as cucumbers are subject to shriveling from loss of moisture. They are waxed to help prevent dehydration.

Eggplant

Eggplants are very sensitive to extremes in temperature. They are held best at 45 to 50°F. Lower temperature will cause chill damage. Bruising is a common problem when careful handling is not employed. Decay will show up as dark brown spots on the surface.

Lettuce

Recommended storage temperature is 32°F. Temperatures higher than 32F will shorten the shelf life of lettuce. Temperature must be maintained as close to 32°F as possible. A storage temperature of 38F will cut the storage life in half. However, lettuce also is damaged easily by freezing; therefore the storage room must be kept above the freezing point at 32°F. Respiration is also a problem with increased temperature. As the temperature rise, so does the respiration rate, which cuts storage life. Leaf lettuce respires at about twice the rate of head lettuce. If temperatures are kept below 35°F, russet spotting is generally kept under control. However, ethylene producing items such as apples, pears and cantaloupes can increase russet spotting. Head lettuce is more susceptible to the problem than other varieties, air circulation is important for lettuce. Lettuce should be stacked so maximum circulation is attained.

Melon

Melons should be stored at 50°F. Melons are quickly affected by extremes of either very hot or very cold temperatures. Careful consideration should be given when storing melons with other commodities. Ethylene gas produced by melons can cause certain vegetables to age and can cause butter to develop an off taste. Some characteristics to look for among the various melons comprise:

- Casaba

Ripeness is indicated by a yellow rind color and slight softening at the blossom end (opposite the stem scar). The flesh should be soft, creamy white, sweet and juicy.

- Crenshaw

Melons are round at the base, coming to a point at the stem end. They will have a gold and green rind that is smooth with no netting and little ribbing. When ripe they will show a softening of rind at the large end, have a golden skin and rich aroma.

Honeydew

These will have a creamy yellow color and velvety surface, but color will vary according to origin. If the outer surface is white with a greenish tint, it is unripe and lacking sugar content. A hard, smooth feel to the rind also indicates the lack of maturity.

Mushroom

The best storage temperature is 32°F. Even under optimum conditions, mushrooms do not keep fresh more than one week after harvest. Under 32°F mushrooms will keep in prime condition for five days, for two days at a temperature of 40F and for one day if temperature is allowed to reach 50°F. Because of their high respiration rate, mushrooms require good air circulation. Mushrooms are comprised mainly of water (90%) and must be protected from dehydration. DO NOT package mushrooms in plastic bags. DO NOT wash mushrooms. Discard large (over 43) over mature pods.

Okra

Store at 40 to 45°F. Discard large, (over 43) over mature pods.

Onion, Bulb

Onions should be stored in a dry cool location. Cured onions can be stored at 32°F to retard sprouting. The best onion will be hard and firm. They should be stored dry. They may require additional drying after harvest in order to store well. 90°F with circulating air enhances drying and curing.

Onion, Green

Green Onions are highly perishable and should be stored at 32°F. Higher temperatures will result in yellowing and leaf decay. Crushed ice is helpful in supplying moisture.

Pepper

Under optimum conditions peppers should not be held for more than 5 to 7 days. The recommended temperature is 45 to 50°F. Peppers are subject to chill damage and pitting if temperatures fall below 45°F. Higher temperatures, above 50°F will encourage ripening or development of red coloring and rapid development of decay.

Potato

Store in a cool, dark location. Potatoes should not be refrigerated below 45°F or they will convert starch to sugar resulting in a too sweet flavor. Light is also a factor as even small amounts of light will cause potatoes to turn green. Potatoes, which have been stored below 42°F, can be conditioned by storing a week to 10 days in temperatures of 50°F or slightly higher. This will convert the sugar back to natural starch and restore the balance that gives potatoes their fresh flavor.

Radish

Radishes will store best at 32°F. As with carrots, topped radishes will last longer than bunched radishes.

Spinach Green

Store at 32°F with high humidity.

Squash

For storage periods of less than four days hold summer squash at 32 to 40°F and then use immediately when removed. If longer than four to five days, temperatures should be 45 to 50°F. Storage life under optimum conditions usually is two weeks for fair to good quality.

Sweet Corn

Sweet corn requires immediate cooling after harvest. Use of hydrocooling or top icing with immediate refrigeration at 32°F gives best results in preventing conversion of sugars to starch. The loss of sugar is 4 times faster at 70°F than at

32°F. Never bulk store sweet corn unless heavily iced because of its tendency to heat through the pile. Maximum of 4 to 8 day shelf life when stored properly.

Sweet potato

Sweet potatoes should be cured before storing by holding at 80 to 85°F and 90 to 95% relative humidity for 4 to 7 days. After curing, reduce temperature down to 50 to 60°F. At lower temperature, sweet potatoes are subject to chill injury, decay, hard core and pithiness.

Tomato

Tomatoes that are picked mature green and properly handled will ripen into a fully colored product with good flavor. Temperatures recommended for ripening fall between 50 and 70°F. Tomatoes are subject to chill damage at temperatures below 50F and lose flavor quickly. Care in handling is crucial. Bruising hastens spoilage and shortens the shelf life. Tomato quality is dependent upon proper harvesting and handling methods.

Watermelon

Holding watermelons for several days at room temperature can improve flavor and coloring. At 50°F and below the watermelon loses color, develops an off flavor and becomes pitted. At higher temperatures they tend to decay. Immature melons do not ripen to a great extent, once harvested, total sugar does not increase.

Sweating

When fruits or vegetables are removed from a low temperature to a higher one, moisture often condenses from the air on the cool surface of the commodity. This is known as sweating; the higher the relative humidity of the outside air, the more marked it becomes. This is because the dew point of the air is at or above the temperature of the commodity. Sweating should be prevented or minimized whenever possible, particularly with onions and the more tender fruits, because it may favor decay. This does not mean that when products sweat after removal from a refrigerated room they will decay; it does mean that conditions are more favorable for decay than if the surfaces remain dry until consumed. Sweating can be prevented to some extent by allowing fruits and vegetables to warm gradually. Air movement over the product while it is warming is helpful in drying the surface.

STORAGE OF MIXED COMMODITIES

At times it may be necessary to store different products together. This may or may not be safe. With some products there is a cross-transfer of odors. Also volatiles such as ethylene are emitted by some products that may be harmful to others. Combinations that should be avoided in storage rooms are apples or pears with celery, cabbage, carrots, potatoes, or onions; celery with onions or carrots; and citrus fruit with any of the strongly scented vegetables. Pears and apples acquire an unpleasant earthy taste and odor when stored with potatoes. It is recommended that onions, nuts, citrus fruit, and potatoes each be stored separately. Lettuce, carrots and greens are damaged when stored with apples, pears, and many other fruits and some vegetables because of the ethylene that is given off from these products as a natural emanation. Very low concentrations may produce adverse effects. Ethylene also stimulates ripening of many fruits and vegetables. This ripening effect is negligible at low temperatures (e.g., 32°F), but it may have an effect at higher temperatures. For this reason, products such as cucumbers, peppers, and acorn squash, in which retention of green color is desirable and which need to be stored at 45 to 50°F, should not be stored with apples, pears, tomatoes, or other ethylene producing crops such as cantaloupes, Honeydew melons, and tomatoes.

Simple ventilation of storage rooms is the cheapest way of removing excess ethylene providing the outside air is not polluted. Overripe and rotten commodities generate substantial quantities of ethylene and should be removed from storage on a regular basis. Internal combustion engines release some ethylene in their exhaust and should never be left running in an enclosed storage area. Commodities that are particularly sensitive to ethylene should be handled with electric forklifts.

Several commercially available materials either absorb ethylene directly or convert it to inactive compounds. Certain types of activated or brominated charcoal absorb ethylene; however, some cheaper materials utilize potassium permanganate to oxidize ethylene to simple carbon dioxide and water. Purafil is an example of a commercial available absorbent. Manipulation of the storage atmosphere, whether in large storerooms or in small packages, can reduce the detrimental effects of ethylene. In general, reducing oxygen and increasing carbon dioxide serves this purpose and is a commercially acceptable procedure for some products. Other chemicals that directly inhibit either ethylene production or ethylene activity are not cleared for Post harvest use on vegetables. It is sometimes necessary to store or transport different types of commodities in a common area. In order for this to be done successfully, the products must have similar Postharvest requirements. Items shipped in mixed loads must be compatible in terms of the following:

- Temperature, i.e. chilling tolerance or sensitivity
- Relative humidity
- Rate of ethylene production

- Sensitivity to ethylene
- Emanation of volatiles with odors
- Absorption of odors

Compromises may be necessary. Conditions suitable for the most valuable portion of a load should be the main criterion for deciding how to set conditions for the whole lot. Harvesting is the process of gathering a ripe crop from the fields. Reaping is the cutting of grain or pulse for harvest, typically using a scythe, sickle, or reaper. On smaller farms with minimal mechanization, harvesting is the most labor-intensive activity of the growing season. On large mechanized farms, harvesting utilizes the most expensive and sophisticated farm machinery, such as the combine harvester. Process automation has increased the efficiency of both the seeding and harvesting process. Specialized harvesting equipment utilizing conveyor belts to mimic gentle gripping and mass transport replaces the manual task of removing each seedling by hand. The term "harvesting" in general usage may include immediate postharvest handling, including cleaning, sorting, packing, and cooling.

The completion of harvesting marks the end of the growing season, or the growing cycle for a particular crop, and the social importance of this event makes it the focus of seasonal celebrations such as harvest festivals, found in many religions.

Etymology

"Harvest", a noun, came from the Old English word hærf-est (coined before the angles moved from Angeln to the British Islands,) meaning "autumn" (the season), "harvest-time", or "August". (It continues to mean "autumn" in British dialect, and "season of gathering crops" generally.) "The harvest" came to also mean the activity of reaping, gathering, and storing grain and other grown products during the autumn, and also the grain and other grown products themselves. "Harvest" was also verbified: "To harvest" means to reap, gather, and store the harvest (or the crop). People who harvest and equipment that harvests are harvesters; while they do it, they are harvesting.

Crop failure

Crop failure (also known as harvest failure) is an absent or greatly diminished crop yield relative to expectation, caused by the plants being damaged, killed, or destroyed, or affected in some way that they fail to form edible fruit, seeds, or leaves in their expected abundance.

Crop failures can be caused by catastrophic events such as plant disease outbreaks (see e.g. Great Famine (Ireland)), heavy rainfall, volcanic eruptions, storms, floods, or drought, or by slow, cumulative effects of soil degradation, too-high soil salinity, erosion, desertification, usually as results of drainage, overdrafting (for irrigation), overfertilization, or overexploitation. In history, crop failures and subsequent famines have triggered human migration, rural exodus, etc. The proliferation of industrialmonocultures, with their reduction in crop diversity and dependence on heavy use of artificial fertilizers and pesticides, has led to overexploited soils that are nearly incapable of regeneration. Over years, unsustainable farming of land degrades soil fertility and diminishes crop yield. With a steadily growing world population and local overpopulation, even slightly diminishing yields are already the equivalent to a partial harvest failure. Fertilizers obviate the need for soil regeneration in the first place, and international trade prevents local crop failures from developing into famines.

Other uses

Harvesting commonly refers to grain and produce, but also has other uses: fishing and logging are also referred to as harvesting. The term harvest is also used in reference to harvesting grapes for wine. Within the context of irrigation, water harvesting refers to the collection and run-off of rainwater for agricultural or domestic uses. Instead of harvest, the term exploit is also used, as in exploiting fisheries or water resources. Energy harvesting is the process of capturing and storing energy (such as solar power, thermal energy, wind energy, salinity gradients, and kinetic energy) that would otherwise go unexploited. Body harvesting, or cadaver harvesting, is the process of collecting and preparing cadavers for anatomical study. In a similar sense, organ harvesting is the removal of tissues or organs from a donor for purposes of transplanting.

Harvesting or Domestic Harvesting in Canada refers to hunting, fishing, and plant gathering by First Nations, Métis, and Inuit in discussions of aboriginal or treaty rights. For example, in the Gwich'in Comprehensive Land Claim Agreement, "Harvesting means gathering, hunting, trapping or fishing..." Similarly, in the Tlicho Land Claim and Self Government Agreement, "'Harvesting' means, in relation to wildlife, hunting, trapping or fishing and, in relation to plants or trees, gathering or cutting." In a non-agricultural sense, the word "harvesting" is an economic principle which is known as an exit event or liquidity event. For example, if a person or business was to cash out of an ownership position in a company or eliminate their investment in a product, it is known as a harvest strategy.

A harvest festival is an annual celebration that occurs around the time of the main harvest of a given region . Given the differences in climate and crops around the world, harvest festivals can be found at various times at different places. Harvest festivals typically feature feasting, both family and public, with foods that are drawn from crops that come to maturity around the time of the festival. Ample food and freedom from the necessity to work in the fields are two central features of harvest festivals: eating, merriment, contests, music and romance are common features of harvest festivals around the world.

In North America, Canada and the US each have their own Thanksgiving celebrations in October and November.

In Britain, thanks have been given for successful harvests since pagan times. Harvest festival is traditionally held on the Sunday near or of the Harvest Moon. This is the full Moon that occurs closest to the autumn equinox (22 or 23 September). The celebrations on this day usually include singing hymns, praying, and decorating churches with baskets of fruit and food in the festival known as Harvest Festival, Harvest Home, Harvest Thanksgiving or Harvest Festival of Thanksgiving.

In British and English-Caribbean churches, chapels and schools, and some Canadian churches, people bring in produce from the garden, the allotment or farm. The food is often distributed among the poor and senior citizens of the local community, or used to raise funds for the church, or charity. Harvest festivals in Asia include the Chinese Mid-Autumn Festival (-NËyÀ{), one of the most widely spread harvest festivals in the world. In Iran Mehrgan was celebrated in an extravagant style at Persepolis. Not only was it the time for harvest, but it was also the time when the taxes were collected. Visitors from different parts of the Persian Empire brought gifts for the king, all contributing to a lively festival. In India, Makar Sankranti, Thai Pongal, Uttarayana, Lohri, and Magh Bihu or Bhogali Bihu in January, Holi in February–March, Vaisakhi in April and Onam in August–September are a few important harvest festivals. Jews celebrate the week-long harvest festival of Sukkot in the autumn. Observant Jews build a temporary hut or shack called a sukkah, and spend the week living, eating, sleeping, and praying inside of it. A sukkah has only three walls and a semi-open roof to allow the elements to enter. It is reminiscent of the structures Israelite farmers would live in during the harvest, at the end of which they would bring a portion to the Temple in Jerusalem.

History

In 1826 in Scotland, the inventor Reverend Patrick Bell designed (but did not patent) a reaper machine, which used the scissors principle of plant cutting – a principle that is still used today. The Bell machine was pushed by horses. A few Bell machines were available in the United States. In 1835, in the United States, Hiram Moore built and patented the first combine harvester, which was capable of reaping, threshing and winnowing cereal grain. Early versions were pulled by horse, mule or ox teams. In 1835, Moore built a full-scale version with a length of 5.2 m (17 ft), cut width of 4.57 m (15 ft) and by 1839, over 20 ha (50 acres) of crops were harvested. This combine harvester was pulled by 20 horses fully handled by farmhands. By 1860, combine harvesters with a cutting, or swathe, width of several metres were used on American farms. A parallel development in Australia saw the development of the stripper based on the Gallic stripper, by John Ridley and others in South Australia by 1843. The stripper only gathered the heads, leaving the stems in the field. The stripper and later headers had the advantage of less moving parts and only collecting heads, requiring less power to operate. Refinements by Hugh Victor McKay produced a commercially successful combine harvester in 1885, the Sunshine Header-Harvester.

Case harvester, 20+ mule team

John Deere 9870 STS underbelly

Combines, some of them quite large, were drawn by mule or horse teams and used a bullwheel to provide power. Later, steam power was used, and George Stockton Berry integrated the combine with a steam engine using straw to heat the boiler. At the turn of the twentieth century, horse drawn combines were starting to be used on the American plains and Idaho (often pulled by teams of twenty or more horses).

Case IH Axial-Flow combine

In 1911, the Holt Manufacturing Company of California produced a self-propelled harvester. In Australia in 1923, the patented Sunshine Auto Header was one of the first center-feeding self-propelled harvesters. In 1923 in Kansas, the Baldwin brothers and their Gleaner Manufacturing Company patented a self-propelled harvester that included several other modern improvements in grain handling. Both the Gleaner and the Sunshine used Fordson engines; early Gleaners used the entire Fordson chassis and driveline as a platform. In 1929, Alfredo Rotania of Argentina patented a self-propelled harvester.International Harvester started making horse-pulled combines in 1915. At the time, horse powered binders and stand alone threshing machines were more common. In the 1920s, Case Corporation and John Deere made combines and these were starting to be tractor pulled with a second engine aboard the combine to power its workings. The world economic collapse in the 1930s stopped farm equipment purchases, and for this reason, people largely retained the older method of harvesting. A few farms did invest and used Caterpillar tractors to move the outfits.

Tractor-drawn combines (also called pull-type combines) became common after World War II as many farms began to use tractors. An example was the All-Crop Harvester series. These combines used a shaker to separate the grain from the chaff and straw-walkers (grates with small teeth on an eccentric shaft) to eject the straw while retaining the grain. Early tractor-drawn combines were usually powered by a separate gasoline engine, while later models were PTO-powered. These machines either put the harvested crop into bags that were then loaded onto a wagon or truck, or had a small bin that stored the grain until it was transferred to a truck or wagon with an auger. In the U.S., Allis-Chalmers, Massey-Harris, International Harvester, Gleaner Manufacturing Company, John Deere, and Minneapolis Moline are past or present major combine producers. In 1937, the Australian-born Thomas Carroll, working for Massey-Harris in Canada, perfected a self-propelled model and in 1940, a lighter-weight model began to be marketed widely by the company. Lyle Yost invented an auger that would lift grain out of a combine in 1947, making unloading grain much easier. In 1952 Claeys launched the first self-propelled combine harvester in Europe; in 1953, the European manufacturer Claas developed a self-propelled combine harvester named 'Hercules', it could harvest up to 5 tons of wheat a day. This newer kind of combine is still in use and is powered by diesel or gasoline engines. Until the self-cleaning rotary screen was invented in the mid-1960s combine engines suffered from overheating as the chaff spewed out when harvesting small grains would clog radiators, blocking the airflow needed for cooling.

A significant advance in the design of combines was the rotary design. The grain is initially stripped from the stalk by passing along a helical rotor, instead of passing between rasp bars on the outside of a cylinder and a concave. Rotary combines were first introduced by Sperry-New Holland in 1975.

In about the 1980s on-board electronics were introduced to measure threshing efficiency. This new instrumentation allowed operators to get better grain yields by optimizing ground speed and other operating parameters.

A New Holland TX68 with grain platform attached

Combine heads

Combines are equipped with removable heads that are designed for particular crops. The standard header, sometimes called a grain platform, is equipped with a reciprocating knife cutter bar, and features a revolving reel with metal or plastic teeth to cause the cut crop to fall into the auger once it is cut. A variation of the platform, a "flex" platform, is similar but has a cutter bar that can flex over contours and ridges to cut soybeans that have pods close to the ground. A flex head can cut soybeans as well as cereal crops, while a rigid platform is generally used only in cereal grains. Some wheat headers, called "draper" headers, use a fabric or rubber apron instead of a cross auger. Draper headers allow faster feeding than cross augers, leading to higher throughputs due to lower power requirements. On many farms, platform headers are used to cut wheat, instead of separate wheat headers, so as to reduce overall costs. Dummy heads or pick-up headers feature spring-tined pickups, usually attached to a heavy rubber belt. They are used for crops that have already been cut and placed in windrows or swaths. This is particularly useful in northern climates such as western Canada, where swathing kills weeds resulting in a faster dry down. While a grain platform can be used for corn, a specialized corn head is ordinarily used instead. The corn head is equipped with snap rolls that strip the stalk and leaf away from the ear, so that only the ear (and husk) enter the throat. This improves efficiency dramatically, since so much less material must go through the cylinder. The corn head can be recognized by the presence of points between each row. Occasionally rowcrop heads are seen that function like a grain platform, but have points between rows like a corn head. These are used to reduce the amount of weed seed picked up when harvesting small grains.

Self-propelled Gleaner combines could be fitted with special tracks instead of tires or tires with tread measuring almost 10in deep to assist in harvesting rice. Some combines, particularly the pull type, have tires with a diamond tread which prevents sinking in mud. These tracks can fit other combines by having adapter plates made.

Conventional combine

The cut crop is carried up the feeder throat (commonly called the "feederhouse") by a chain and flight elevator, then fed into the threshing mechanism of the combine, consisting of a rotating threshing drum (commonly called the "cylinder"), to which grooved steel bars (rasp bars) are bolted. The rasp bars thresh or separate the grains and chaff from the straw through the action of the cylinder against the concave, a shaped "half drum", also fitted with steel bars and a meshed grill, through which grain, chaff and smaller debris may fall, whereas the straw, being too long, is carried through onto the straw walkers. This action is also allowed because grain is heavier than straw, which causes it to fall rather than "float" across from the cylinder/ concave to the walkers. The drum speed is variably adjustable on most machines, whilst the distance between the drum and concave is finely adjustable fore, aft and together, to achieve optimum separation and output. Manually engaged disawning plates are usually fitted to the concave. These provide extra friction to remove the awns from barley crops. After the primary separation at the cylinder, the clean grain falls through the concave and to the shoe, which contains the chaffer and sieves. The shoe is common to both conventional combines and rotary combines.

Hillside leveling

In the Palouse region of the Pacific Northwest of the United States the combine is retrofitted with a hydraulic hillside leveling system. This allows the combine to harvest the steep but fertile soil in the region. Hillsides can be as steep as a 50% slope. Gleaner, IH and Case IH, John Deere, and others all have made combines with this hillside leveling system, and local machine shops have fabricated them as an aftermarket add-on.

The first leveling technology was developed by Holt Co., a California firm, in 1891. Modern leveling came into being with the invention and patent of a level sensitive mercury switch system invented by Raymond Alvah Hanson in 1946. Raymond's son, Raymond, Jr., produced leveling systems exclusively for John Deere combines until 1995 as R. A. Hanson Company, Inc. In 1995, his son, Richard, purchased the company from his father and renamed it RAHCO International, Inc. In March 2011, the company was renamed Hanson Worldwide, LLC. Production continues to this day. Hillside leveling has several advantages. Primary among them is an increased threshing efficiency on hillsides. Without leveling, grain and chaff slide to one side of separator and come through the machine in a large ball rather than being separated, dumping large amounts of grain on the ground. By keeping the machinery level, the straw-walker is able to operate more efficiently, making for more efficient threshing. IH produced the 453 combine which leveled both side-to-side and front-to-back, enabling efficient threshing whether on a hillside or climbing a hill head on. Secondarily, leveling changes a combine's center of gravity relative to the hill and allows the combine to harvest along the contour of a hill without tipping, a danger on the steeper slopes of the region; it is not uncommon for combines to roll over on extremely steep hills. Newer leveling systems do not have as much tilt as the older ones. A John Deere 9600 combine equipped with a Rahco hillside conversion kit will level to 44%, while the newer STS combines will only go to 35%. These modern combines use the rotary grain separator which makes leveling less critical. Most combines on the Palouse have dual drive wheels on each side to stabilize them. A leveling system was developed in Europe by the Italian combine manufacturer Laverda which still produces it today.

Sidehill leveling

Sidehill combines are very similar to hillside combines in that they level the combine to the ground so that the threshing can be efficiently conducted; however, they have some very distinct differences. Modern hillside combines level around 35% on average, while older machines were closer to 50%. Sidehill combines only level to 18%. They are sparsely used in the Palouse region. Rather, they are used on the gentle rolling slopes of the midwest. Sidehill combines are much more mass-produced than their hillside counterparts. The height of a sidehill machine is the same height as a level-land combine. Hillside combines have added steel that sets them up approximately 2–5 feet higher than a level-land combine and provide a smooth ride

Maintaining threshing speed

Another technology that is sometimes used on combines is a continuously variable transmission. This allows the ground speed of the machine to be varied while maintaining a constant engine and threshing speed. It is desirable to keep the threshing speed constant since the machine will typically have been adjusted to operate best at a certain speed. Self-propelled combines started with standard manual transmissions that provided one speed based on input rpm. Deficiencies were noted and in the early 1950s combines were equipped with what John Deere called the "Variable Speed Drive". This was simply a variable width sheave controlled by spring and hydraulic pressures. This sheave was attached to the input shaft of the transmission. A standard 4 speed manual transmission was still used in this drive system. The operator would select a gear, typically 3rd. An extra control was provided to the operator to allow him to speed up and slow down the machine within the limits provided by the variable speed drive system. By decreasing the width of the sheave on the input shaft of the transmission, the belt would ride higher in the groove. This slowed the rotating speed on the input shaft of the transmission, thus slowing the ground speed for that gear. A clutch was still provided to allow the operator to stop the machine and change transmission gears.

Later, as hydraulic technology improved, hydrostatic transmissions were introduced by Versatile Mfg for use on swathers but later this technology was applied to combines as well. This drive retained the 4 speed manual transmission as before, but this time used a system of hydraulic pumps and motors to drive the input shaft of the transmission. This system is called a Hydrostatic drive system. The engine turns the hydraulic pump capable of pressures up to 4,000 psi (30 MPa). This pressure is then directed to the hydraulic motor that is connected to the input shaft of the transmission. The operator is provided with a lever in the cab that allows for the control of the hydraulic motor's ability to use the energy provided by the pump. By adjusting the swash plate in the motor, the stroke of its pistons are changed. If the swash plate is set to neutral, the pistons do not move in their bores and no rotation is allowed, thus the machine does not move. By moving the lever, the swash plate moves its attached pistons forward, thus allowing them to move within the bore and causing the motor to turn. This provides an infinitely variable speed control from 0 ground speed to what ever the maximum speed is allowed by the gear selection of the transmission. The standard clutch was removed from this drive system as it was no longer needed. Most if not all modern combines are equipped with hydrostatic drives. These are larger versions of the same system used in consumer and commercial lawn mowers that most are familiar with today. In fact, it was the down-sizing of the combine drive system that placed these drive systems into mowers and other machines.

The threshing process

Despite great advances in mechanics and computer control, the basic operation of the combine harvester has remained unchanged almost since it was invented.

First, the header, described above, cuts the crop and feeds it into the threshing cylinder. This consists of a series of horizontal rasp bars fixed across the path of the crop and in the shape of a quarter cylinder. Moving rasp bars or rub bars pull the crop through concaved grates that separate the grain and chaff from the straw. The grain heads fall through the fixed concaves. What happens next is dependent on the type of combine in question. In most modern combines, the grain is transported to the shoe by a set of 2, 3, or 4 (possibly more on the largest machines) augers, set parallel or semi-parallel to the rotor on axial mounted rotors and perpendicular on "Axial Flow" combines. In older Gleaner machines, these augers were not present. Those combines are unique in that the cylinder and concave is set inside feederhouse instead of in the machine directly behind the feederhouse. Consequently, the material was moved by a "raddle chain" from underneath the concave to the walkers. The clean grain fell between the raddle and the walkers onto the shoe, while the straw, being longer and lighter, floated across onto the walkers to be expelled. On most other older machines, the cylinder was placed higher and farther back in the machine, and the grain moved to the shoe by falling down a "clean grain pan", and the straw "floated" across the concaves to the back of the walkers.

Since the Sperry-New Holland TR70 Twin-Rotor Combine came out in 1975, most manufacturers have combines with rotors in place of conventional cylinders. However, makers have now returned to the market with conventional models alongside their rotary line-up. A rotor is a long, longitudinally mounted rotating cylinder with plates similar to rub bars (except for in the above-mentioned Gleaner rotaries). There are usually two sieves, one above the other. The sieves and basically a metal frame, that has many rows of "fingers" set reasonably close together. The angle of the fingers is adjustable as to change the clearance and control the size of material passing through. The top is set with more clearance than the bottom as to allow a gradual cleaning action. Setting the concave clearance, fan speed, and sieve size is critical to ensure that the crop is threshed properly, the grain is clean of debris, and that all of the grain entering the machine reaches the grain tank or 'hopper'. (Observe, for example, that when travelling uphill the fan speed must be reduced to account for the shallower gradient of the sieves.)

Heavy material, e.g., unthreshed heads, fall off the front of the sieves and are returned to the concave for re-threshing.

The straw walkers are located above the sieves, and also have holes in them. Any grain remaining attached to the straw is shaken off and falls onto the top sieve.

When the straw reaches the end of the walkers it falls out the rear of the combine. It can then be baled for cattle bedding or spread by two rotating straw spreaders with rubber arms. Most modern combines are equipped with a straw spreader.

Rather than immediately falling out the rear of the combine at the end of the walkers, there are models of combine harvesters from Eastern Europe and Russia (e.g. Agromash Yenisei 1200 1 HM, etc.) that have a "straw catchers" at the end of the walkers that temporarily hold the straw and then once full, deposit it in a stack for easy gathering.

CONVENTIONAL AND ROTARY DESIGNS

Tr85

For some time, combine harvesters used the conventional design, which used a rotating cylinder at the front-end which knocked the seeds out of the heads, and then used the rest of the machine to separate the straw from the chaff, and the chaff from the grain. The TR70 from Sperry-New Holland was brought out in 1975 as the first rotary combine. Other manufacturers soon followed, International Harvester with their 'Axial Flow' in 1977 and Gleaner with their N6 in 1979.

In the decades before the widespread adoption of the rotary combine in the late seventies, several inventors had pioneered designs which relied more on centrifugal force for grain separation and less on gravity alone. By the early eighties, most major manufacturers had settled on a "walkerless" design with much larger threshing cylinders to do most of the work. Advantages were faster grain harvesting and gentler treatment of fragile seeds, which were often cracked by the faster rotational speeds of conventional combine threshing cylinders.

It was the disadvantages of the rotary combine (increased power requirements and over-pulverization of the straw by-product) which prompted a resurgence of conventional combines in the late nineties. Perhaps overlooked but nonetheless true, when the large engines that powered the rotary machines were employed in conventional machines, the two types of machines delivered similar production capacities. Also, research was beginning to show that incorporating above-ground crop residue (straw) into the soil is less useful for rebuilding soil fertility than previously believed. This meant that working pulverized straw into the soil became more of a hindrance than a benefit. An increase in feedlot beef production also created a higher demand for straw as fodder. Conventional combines, which use straw walkers, preserve the quality of straw and allow it to be baled and removed from the field.

Combine fires

Grain combine fires are responsible for millions of dollars of loss each year. Fires usually start near the engine where dust and dry crop debris accumulate. Fires can also start when heat is introduced by bearings or gearboxes that have failed. From 1984 to 2000, 695 major grain combine fires were reported to U.S. local fire departments. Dragging chains to reduce static electricity was one method employed for preventing harvester fires, but the role of static electricity linked to causing harvester fires is yet to be established. The application of the appropriate synthetic greases will reduce the friction experienced at crucial points, i.e. chains, sprockets and gear boxes compared to petroleum based lubricants. Engines with synthetic lubricants will also remain significantly cooler during operation.

Information and communications technology in agriculture

Information and communication technology in agriculture (ICT in agriculture), also known as e-agriculture, focuses on the enhancement of agricultural and rural development through improved information and communication processes. More specifically, e-agriculture involves the conceptualization, design, development, evaluation and application of innovative ways to use information and communication technologies (ICTs) in the rural domain, with a primary focus on agriculture. ICT includes devices, networks, mobiles, services and applications; these range from innovative Internet-era technologies and sensors to other pre-existing aids such as fixed telephones, televisions, radios and satellites. Provisions of standards, norms, methodologies, and tools as well as development of individual and institutional capacities, and policy support are all key components of e-agriculture. Many ICT in agriculture or e-agriculture interventions have been developed and tested around the world to help agriculturists improve their livelihoods through increased agricultural productivity and income, or by reducting risks. Some useful resources for learning about e-agriculture in practice are the World Bank's e-sourcebook ICT in agriculture – connecting smallholder farmers to knowledge, networks and institutions (2011), ICT uses for inclusive value chains (2013), ICT uses for inclusive value chains (2013) and Success stories on information and communication technologies for agriculture and rural development have documented many cases of use of ICT in agriculture.

Wireless technologies

Wireless technologies have numerous applications in agriculture. One major usage is the simplification of closed-circuit television camera systems; the use of wireless communications eliminates the need for the installation of coaxial cables.

Global Positioning System (GPS)

In agriculture, the use of the Global Positioning System provides benefits in geo-fencing, map-making and surveying. GPS receivers dropped in price over the years, making it more popular for civilian use. With the use of GPS, civilians can produce simple yet highly accurate digitized map without the help of a professional cartographer. In Kenya, for example, the solution to prevent an elephant bull from wandering into farms and destroying precious crops was to tag the elephant with a device that sends a text message when it crosses a geo-fence. Using the technology of SMS and GPS, the elephant can roam freely and the authorities are alerted whenever it is near the farm.

Geographic information systems

Geographic information systems, or GiS, are extensively used in agriculture, especially in precision farming. Land is mapped digitally, and pertinent geodetic data such as topography and contours are combined with other statistical data for easier analysis of the soil. GIS is used in decision making such as what to plant and where to plant using historical data and sampling.

Computer-controlled devices (automated systems)

Automatic milking systems are computer controlled stand alone systems that milk the dairy cattle without human labor. The complete automation of the milking process is controlled by an agricultural robot, a complex herd management software, and specialized computers. Automatic milking eliminates the farmer from the actual milking process, allowing for more time for supervision of the farm and the herd. Farmers can also improve herd management by using the data gathered by the computer. By analyzing the effect of various animal feeds on milk yield, farmers may adjust accordingly to obtain optimal milk yields. Since the data is available down to individual level, each cow may be tracked and examined, and the farmer may be alerted when there are unusual changes that could mean sickness or injuries.

Smartphone mobile apps in agriculture

The use of mobile technologies as a tool of intervention in agriculture is becoming increasingly popular. Smartphone penetration enhances the multi-dimensional positive impact on sustainable poverty reduction and identify accessibility as the main challenge in harnessing the full potential (Silarszky et al., 2008) in agricultural space. The reach of smartphone even in rural areas extended the ICT services beyond simple voice or text messages. Several smartphone apps are available for agriculture, horticulture, animal husbandry and farm machinery.

Recognition and application

E-agriculture is one of the action lines identified in the declaration and plan of action of the World Summit on the Information Society (WSIS). The "Tunis Agenda for the Information Society," published on 18 November 2005 and emphasizes the leading facilitating roles that UN agencies need to play in the implementation of the Geneva Plan of Action. The Food and Agriculture Organization of the United Nations (FAO) has been assigned the responsibility of organizing activities related to the action line under C.7 ICT Applications on E-Agriculture.

Many ICT interventions have been developed and tested around the world, with varied degrees of success, to help agriculturists improve their livelihoods through increased agricultural productivity and incomes, and reduction in risks. Some useful resources for learning about e-agriculture in practice are the World Bank's e-sourcebook ICT in agriculture – connecting smallholder farmers to knowledge, networks and institutions (2011), ICT uses for inclusive value chains (2013), ICT uses for inclusive value chains (2013) and Success stories on information and communication technologies for agriculture and rural development have documented many cases of use of ICT in agriculture.

The FAO-ITU E-agriculture Strategy Guide was developed by the Food and Agriculture Organization (FAO) and the International Telecommunication Union (ITU) with support from partners including the Technical Centre for Agricultural and Rural Cooperation (CTA) as a framework for countries in developing their national e-agriculture strategy/masterplan.

Some of the countries who are using the FAO-ITU E-agriculture Strategy Guide to develop their national e-agriculture strategy are Bhutan, Sri Lanka, Papua New Guinea, Philippines, Fiji and Vanuatu. The guide provides a framework to engage broader stakeholders in the development of national e-agriculture strategy.

History

In 2008, the United Nations referred to e-agriculture as "an emerging field", with the expectation that its scope would change and evolve as our understanding of the area grows.

ICT in support of Rural Poverty Elimination and Food Security

In August 2003, the Overseas Development Institute (ODI), the UK Department for International Development (DFID) and the United Nations Food and Agricultural Organization (FAO) joined together in a collaborative research project to look at bringing together livelihoods thinking with concepts from information and communication for development, in order to improve understanding of the role and importance of information and communication in support of rural livelihoods.

The policy recommendations included:

- Building on existing systems, while encouraging integration of different technologies and information sharing
- Determining who should pay, through consensus and based on a thorough analysis of the costs
- Ensuring equitable access to marginalised groups and those in the agricultural sector
- Promoting localised content, with decentralised and locally owned processes
- Building capacity, through provision of training packages and maintaining a choice of information sources
- Using realistic technologies, that are suitable within the existing infrastructure
- Building knowledge partnerships to ensure that knowledge gaps are filled and a two-way flow of information allows knowledge to originate from all levels of the network and community.

The importance of ICT is also recognized in the 8th Millennium Development Goal, with the target to "...make available the benefits of new technologies, especially information and communications technologies (ICTs)" to the fight against poverty.

WSIS process

E-agriculture is one of the action lines identified in the declaration and plan of action (2003) of the World Summit on the Information Society (WSIS). The "Tunis Agenda for the Information Society", published on 18 November 2005, emphasizes the leading facilitating roles that UN agencies need to play in the implementation of the Geneva Plan of Action.

FAO hosted the first e-agriculture workshop in June 2006, bringing together representatives of leading development organizations involved in agriculture. The meeting served to initiate development of an effective process to engage as wide a range of stakeholders involved in e-agriculture, and resulted in the formation of the e-Agriculture Community, a community of practice. The e-Agriculture Community's Founding Partners include: Consultative Group on International Agricultural Research (CGIAR); Technical Centre for Agriculture and Rural Development (CTA); FAO; Global Alliance for Information and Communication Technologies and Development (GAID); Global Forum on Agricultural Research (GFAR); Global Knowledge Partnership (GKP); Gesellschaft fur Technische Zusammenarbeit (now called Deutsche Gesellschaft für Internationale Zusammenarbeit, GIZ); International Association of Agricultural Information Specialists (IAALD); Inter-American Institute for Cooperation on Agriculture (IICA); International Fund for Agricultural Development (IFAD); International Centre for Communication for Development (IICD); United States National Agricultural Library (NAL); United Nations Department of Economic and Social Affairs (UNDESA); the World Bank. An agricultural cooperative, also known as a farmers' co-op, is a cooperative where farmers pool their resources in certain areas of activity. A broad typology of agricultural cooperatives distinguishes between 'agricultural service cooperatives', which provide various services to their individually farming members, and 'agricultural production cooperatives', where production resources (land, machinery) are pooled and members farm jointly. Examples of agricultural production cooperatives include collective farms in former socialist countries, the kibbutzim in Israel, collectively governed community shared agriculture, Longo Mai co-operatives and Nicaraguan production co-operatives.

The default meaning of 'agricultural cooperative' in English is usually an agricultural 'service' cooperative, which is the numerically dominant form in the world. There are two primary types of agricultural service cooperatives, 'supply cooperative' and 'marketing cooperative'. Supply cooperatives supply their members with inputs for agricultural production, including seeds, fertilizers, fuel, and machinery services. Marketing cooperatives are established by farmers to undertake transportation, packaging, distribution, and marketing of farm products (both crop and livestock). Farmers also widely rely on credit cooperatives as a source of financing for both working capital and investments.

Purpose

Cooperatives as a form of business organization are distinct from the more common investor-owned firms (IOFs). Both are organized as corporations, but IOFs pursue profit maximization objectives, whereas cooperatives strive to maximize the benefits they generate for their members (which usually involves zero-profit operation). Agricultural cooperatives are therefore created in situations where farmers cannot obtain essential services from IOFs (because the provision of these services is judged to be unprofitable by the IOFs), or when IOFs provide the services at disadvantageous terms to the farmers (i.e., the services are available, but the profit-motivated prices are too high for the farmers). The former situations are characterized in economic theory as market failure or missing services motive. The latter drive the creation of cooperatives as a competitive yardstick or as a means of allowing farmers to build countervailing market power to oppose the IOFs. The concept of competitive yardstick implies that farmers, faced with unsatisfactory performance by IOFs, may form a cooperative firm whose purpose is to force the IOFs, through competition, to improve their service to farmers.

Headquarters of Hokuren Federation of Agricultural Cooperatives in Sapporo, Japan A practical motivation for the creation of agricultural cooperatives is related to the ability of farmers to pool production and/or resources. In many situations within agriculture, it is simply too expensive for farmers to manufacture products or undertake a service. Cooperatives provide a method for farmers to join together in an 'association', through which a group of farmers can acquire a better outcome, typically financial, than by going alone. This approach is aligned to the concept of economies of scale and can also be related as a form of economic synergy, where "two or more agents working together to produce a result not obtainable by any of the agents independently". While it may seem reasonable to conclude that larger the cooperative the better, this is not necessarily true. Cooperatives exist across a broad membership base, with some cooperatives having fewer than 20 members while others can have over 10,000. While the economic benefits are a strong driver in forming cooperatives, it is not the sole consideration. In fact, it is possible for the economic benefits from a cooperative to be replicated in other organisational forms, such as an IOF. An important strength of a cooperative for the farmer is that they retain the governance of the association, thereby ensuring they have ultimate ownership and control. This ensures that the profit reimbursement (either through the dividend payout or rebate) is shared only amongst the farmer members, rather than shareholders as in an IOF.

In agriculture, there are broadly three types of cooperatives: a machinery pool, a manufacturing/marketing cooperative, and a credit union.

- Machinery pool: A family farm may be too small to justify the purchase of expensive farm machinery, which may be only used irregularly, say only during harvest; instead local farmers may get together to form a machinery pool that purchases the necessary equipment for all the members to use.
- Manufacturing/marketing cooperative: A farm does not always have the means of transportation necessary for delivering its produce to the market, or else the small volume of its production may put it in an unfavorable negotiating position with respect to intermediaries and wholesalers; a cooperative will act as an integrator, collecting the output from members, sometimes undertaking manufacturing, and delivering it in large aggregated quantities downstream through the marketing channels.
- Credit Union: Farmers, especially in developing countries, can be charged relatively high interest rates by commercial banks, or credit may not even be available for farmers to access. When providing loans, these banks are often mindful of high transaction costs on small loans, or may refuse credit altogether due to lack of collateral – something very acute in developing countries. To provide a source of credit, farmers can group together funds that can be loaned out to members. Alternatively, the credit union can raise loans at better rates from commercial banks due to the cooperative having a larger associative size than an individual farmer. Often members of a credit union will provide mutual or peer-pressure guarantees for repayment of loans. In some instances, manufacturing/marketing cooperatives may have credit unions as part of their broader business. Such an approach allows farmers to have a more direct access to critical farm inputs, such as seeds and implements. The loans for these inputs are repaid when the farmer sends produce to the manufacturing/marketing cooperative.

Origins

The first agricultural cooperatives were created in Europe in the seventeenth century in the Military Frontier, where the wives and children of the border guards lived together in organized agricultural cooperatives next to a funfair and a public bath. The first civil agricultural cooperatives were created also in Europe in the second half of the nineteenth century. They spread later to North America and the other continents. They have become one of the tools of agricultural development in emerging countries. Farmers also cooperated to form mutual farm insurance societies.

Also related are rural credit unions. They were created in the same periods, with the initial purpose of offering farm loans. Some became universal banks such as Crédit Agricole or Rabobank.

Supply cooperatives

Agricultural supply cooperatives aggregate purchases, storage, and distribution of farm inputs for their members. By taking advantage of volume discounts and utilizing other economies of scale, supply cooperatives bring down the cost of the inputs that the members purchase from the cooperative compared with direct purchases from commercial suppliers.

CUSTOM HARVESTING

In agriculture, custom harvesting or custom combining is the business of harvesting of crops for others. Custom harvesters usually own their own combines and work for the same farms every harvest season. Custom harvesting relieves farmers from having to invest capital in expensive equipment while at the same time maximizing the machinery's use.

The custom harvesting industry has its roots in the mid-twentieth century. Before the invention of the combine harvester, farmers usually owned their own harvesting machinery and worked in tandem with migrant workers, who would bring their own threshing equipment. As combines became more and more widespread, the demand for migrant labor decreased. Custom harvesters, owning their own combines, existed beginning in the 1920s, albeit on a small scale. However, World War II caused a labor and materials shortage, and the custom harvesting industry experienced a great growth. In North America, the area from Texas to Saskatchewan was heavily traversed by custom harvesters. In 1942, there were approximately 500 such operators in this range, but by 1947 that number had grown to 8,000 in Kansas alone. Over the years, the industry transformed from one being run solely by men into family-run businesses. Harvesters would pass their routes and clients down to their children, thus turning custom combining into family tradition.

Custom combiners charge for their services in several ways: a flat per-unit of land price, a per-bale price, or a percentage of the crop. One of the biggest problems for farmers who hire custom harvesters is the availability of labor and the timing of the harvest. If the harvester cannot be available right when the crops are ready, there is the risk of low harvest (if too early) or spoilage (if too late).

The number of custom harvesters has dropped in modern times due to a number of reasons. For one, the consolidation of farms into large-scale operations has allowed them to operate on economies of scale such that they do not need to rely on contractors. Competition is another reason. In China, custom harvesting became such a popular profession in the late 20th century that too many people entered the industry and prices became too low to generate a profit.

Energy harvesting (also known as power harvesting or energy scavenging or ambient power) is the process by which energy is derived from external sources (e.g., solar power, thermal energy, wind energy, salinity gradients, and kinetic energy, also known as ambient energy), captured, and stored for small, wireless autonomous devices, like those used in wearable electronics and wireless sensor networks.

Energy harvesters provide a very small amount of power for low-energy electronics. While the input fuel to some large-scale generation costs resources (oil, coal, etc.), the energy source for energy harvesters is present as ambient background. For example, temperature gradients exist from the operation of a combustion engine and in urban areas, there is a large amount of electromagnetic energy in the environment because of radio and television broadcasting.

One of the earliest applications of ambient power collected from ambient electromagnetic radiation (EMR) is the crystal radio.

The principles of energy harvesting from ambient EMR can be demonstrated with basic components.

Operation

Energy harvesting devices converting ambient energy into electrical energy have attracted much interest in both the military and commercial sectors. Some systems convert motion, such as that of ocean waves, into electricity to be used by oceanographic monitoring sensors for autonomous operation. Future applications may include high power output devices (or arrays of such devices) deployed at remote locations to serve as reliable power stations for large systems. Another application is in wearable electronics, where energy harvesting devices can power or recharge cellphones, mobile computers, radio communication equipment, etc.

All of these devices must be sufficiently robust to endure long-term exposure to hostile environments and have a broad range of dynamic sensitivity to exploit the entire spectrum of wave motions.

Accumulating energy

Energy can also be harvested to power small autonomous sensors such as those developed using MEMS technology. These systems are often very small and require little power, but their applications are limited by the reliance on battery power. Scavenging energy from ambient vibrations, wind, heat or light could enable smart sensors to be functional indefinitely.

Typical power densities available from energy harvesting devices are highly dependent upon the specific application (affecting the generator's size) and the design itself of the harvesting generator. In general, for motion powered devices, typical values are a few $\mu W/cm^3$ for human body powered applications and hundreds of $\mu W/cm^3$ for generators powered from machinery. Most energy scavenging devices for wearable electronics generate very little power.

Storage of power

In general, energy can be stored in a capacitor, super capacitor, or battery. Capacitors are used when the application needs to provide huge energy spikes. Batteries leak less energy and are therefore used when the device needs to provide a steady flow of energy. Compared to batteries, super capacitors have virtually unlimited charge-discharge cycles and can therefore operate forever enabling a maintenance-free operation in IoT and wireless sensor devices .

Use of the power

Current interest in low power energy harvesting is for independent sensor networks. In these applications an energy harvesting scheme puts power stored into a capacitor then boosted/regulated to a second storage capacitor or battery for the use in the microprocessor or in the data transmission. The power is usually used in a sensor application and the data stored or is transmitted possibly through a wireless method.

Motivation

The history of energy harvesting dates back to the windmill and the waterwheel. People have searched for ways to store the energy from heat and vibrations for many decades. One driving force behind the search for new energy harvesting devices is the desire to power sensor networks and mobile devices without batteries. Energy harvesting is also motivated by a desire to address the issue of climate change and global warming.

Devices

There are many small-scale energy sources that generally cannot be scaled up to industrial size:

- Some wristwatches are powered by kinetic energy (called automatic watches), in this case movement of the arm is used. The arm movement causes winding of its mainspring. A newer design introduced by Seiko ("Kinetic") uses movement of a magnet in the electromagnetic generator instead to power the quartz movement. The motion provides a rate of change of flux, which results in some induced emf on the coils. The concept is related to Faraday's Law.
- Photovoltaics is a method of generating electrical power by converting solar radiation (both indoors and outdoors) into direct current electricity using semiconductors that exhibit the photovoltaic effect. Photovoltaic power generation employs solar panels composed of a number of cells containing a photovoltaic material. Note that photovoltaics have been scaled up to industrial size and that large solar farms exist.
- Thermoelectric generators (TEGs) consist of the junction of two dissimilar materials and the presence of a thermal gradient. Large voltage outputs are possible by connecting many junctions electrically in series and thermally in parallel. Typical performance is 100-300 ìV/K per junction. These can be utilized to capture mW.s of energy from industrial equipment, structures, and even the human body. They are typically coupled with heat sinks to improve temperature gradient.
- Micro wind turbine are used to harvest wind energy readily available in the environment in the form of kinetic energy to power the low power electronic devices such as wireless sensor nodes. When air flows across the blades of the turbine, a net pressure difference is developed between the wind speeds above and below the blades. This will result in a lift force generated which in turn rotate the blades. Similar to photovoltaics, wind farms have been constructed on an industrial scale and are being used to generate substantial amounts of electrical energy.
- Piezoelectric crystals or fibers generate a small voltage whenever they are mechanically deformed. Vibration from engines can stimulate piezoelectric materials, as can the heel of a shoe, or the pushing of a button.

- Special antennas can collect energy from stray radio waves, this can also be done with a Rectenna and theoretically at even higher frequency EM radiation with a Nantenna.
- Power from keys pressed during use of a portable electronic device or remote controller, using magnet and coil or piezoelectric energy converters, may be used to help power the device.

Mechanized agriculture

The seed drill is a mechanical device for spacing and planting seed at the appropriate depth. It originated in ancient China before the 1st century BC. Saving seed was extremely important at a time when yields were measured in terms of seeds harvested per seed planted, which was typically between 3 and 5. The seed drill also saved planting labor. Most importantly, the seed drill meant crops were grown in rows, which reduced competition of plants and increase yields. It was reinvented in 16th century Europe based on verbal descriptions and crude drawings brought back from China.Jethro Tull patented a version in 1700; however, it was expensive and unreliable. Reliable seed drills appeared in the mid 19th century.

Since the beginning of agriculture threshing was done by hand with a flail, requiring a great deal of labor. The threshing machine (ca. 1794) simplified the operation and allowed it to use animal power. By the 1860s threshing machines were widely introduced and ultimately displaced as much as a quarter of agricultural labor. In Europe, many of the displaced workers were driven to the brink of starvation.

Before c. 1790 a worker could harvest 1/4 acre per day with a scythe. In the early 1800s the grain cradle was introduced, significantly increasing the productivity of hand labor. It was estimated that each of Cyrus McCormick's horse pulled reapers (Ptd. 1834) freed up five men for military service in the U.S. Civil War. By 1890 two men and two horses could cut, rake and bind 20 acres of wheat per day. In the 1880s the reaper and threshing machine were combined into the combine harvester. These machines required large teams of horses or mules to pull. Over the entire 19th century the output per man hour for producing wheat rose by about 500% and for corn about 250%. Farm machinery and higher crop yields reduced the labor to produce 100 bushels of corn from 35 to 40 hours in 1900 to 2 hours 45 minutes in 1999. The conversion of agricultural mechanization to internal combustion power began after 1915. The horse population began to decline in the 1920s after the conversion of agriculture and transportation to internal combustion. In addition to saving labor, this freed up much land previously used for supporting draft animals.

6

GENETIC RESOURCE OF HORTICULTURE CROP

Plant biodiversity has a great relevance to the evolution of mankind from hunters and gatherers to organized agriculture. In fact, Bible mentions about the "Forbidden fruit" and "Garden of Eden" when God created the life forms. They were indications of the importance of the Horticulture. The origin of Botanic gardens started since then. Sacred groves are part of this evolution. India is home to a large number of agricultural crops (including horticulture). It is considered to be the homeland of 167 important plant species of cereals, millets, fruits, condiments, vegetables, pulses, fibre crops and oilseeds, and 114 breeds of domesticated animals. About 4,900 species of flowering plants are endemic to the country. These are distributed among 141 genera belonging to 47 families.

India is the centre of diversity for many Horticultural crops, which have all kinds of endemic varieties, alleles and even, Linnean species. India is known as land of spices, being the origin of two important spices, black pepper and cardamom and also for ginger, turmeric, Garcinia and Myristica with maximum diversity. Rich diversity also occurs in India for medicinal and aromatic plants and traditional knowledge associated with their uses, particularly in the Western Ghats and North-Eastern region. Besides, Indian subcontinent is rich repository for ornamental trees, shrubs, climbers, herbs and succulents. Wild relatives are also available in plenty in the rain forest habitats. The Horticultural germplasm is being conserved both under in situ and ex-situ gene banks. The horticulture crops germplasm of over 36000 are being conserved under different ICAR institutes in India covering nearly about 130 horticultural crops The advent of commercialization and development of urban based Horticulture led to the loss of genetic materials such as land races and Heirlooms. Heirloom varieties can be of immense value to a researcher as they can be used in the breeding programme and can be used by farmers for other value added products and help in getting better income when the commercial varieties are not in fruiting. The potential of vegetable genetic resource (VGR) is tremendous, enabling us to meet with future challenges such as food and nutritional security. This is achievable through identification, collection, conservation, documentation and utilization of these VGRs. The development and adoption of improved varieties has led to replacement of innumerable landraces and local crop cultivars of different agri-horticultural crops in general and vegetables in particular. Harmonization of activities relating to plant genetic resource management at the global and national level has improved the scope of VGR utilization.

Precise documentation on the strength of vegetable genetic resources of a region is a must for agri-biodiversity assessment. India is rich in vegetable genetic resource diversity of indigenous and exotic origin. The information on VGR of Indian region can be made available through passport data, conservation data, catalogues on characterization/evaluation and information on local use of vegetables to further strengthen these activities. In this chapter the vegetable genetic resources particularly the important cultivated species from the Indian region along with regionally important crops, less-known types, potential species and wild relatives were discussed in light of efforts made at national level. Future targets are highlighted on the basis of gap areas identified in VGR management.

Disadvantages and Advantages of Genetically Modified Crops

Nearly most of the crops that are grown in the US have been touched by science. Having genes that have been manipulated, genetically modified crops are becoming more and more popular in the country, with people liking the fact that they provide certain perks, such as larger yields, insect resistance and faster growing traits. However, they also do not come with criticism, where many people believe that research on the technology is not sufficient. In order to get a better understanding of genetically modified crops, it is best to take a look at their pros and cons.

List of Pros of Genetically Modified Crops

1. They are a technology that has already been going on for many years.- Genetic modification and engineering is nothing new, where people have been already manipulating crops and food for many years, using

methods that range from classical selection to fermentation. It is just the latest form of biotechnology that is using the most precise method known. Basically, creating these crops is different from the traditional methods of plant breeding, as it transfers genes from one species to another, crossing biological barriers.

2. They are mostly believed to be safe.-Proponents say that there are actually no inherent differences between food items produced from natural crops and those from genetically modified crops, similarly containing DNA that comprised of nucleotides. This technology works by moving a piece of DNA from one plant breed into another, without introducing a foreign substance and with the new gene merely prompting the modified crop to show a desired trait. Also companies wishing to release a product or seed from a genetically modified crop are required the safety of these things. And if a certain product was created from an organism that contains a known allergen, it is required to undergo some tests for safety.
3. They have improved quality and taste, as well as increased nutrients.-Genetic modification in crops allows flavors of its products to be improved. For example, pepper can become spicier, corn can become sweeter and bitter gourd would become palatable without losing their original nutrients. Better yet, this technology makes it possible to enhance the vitamin and mineral content of food, providing benefits to consumers. This advantage is especially important to people in developing countries, who do not always have access to essential nutritional resources.
4. They help ease the environmental impact by human activities.-With the way it is practiced today, agriculture sometimes damages the environment, like any other human activity. Now, genetically modified crops can help ease such negative impact by creating plants that are insect resistant, allowing farmers to dramatically reduce the use of insecticides. Also, the technology can create next-generation seeds that will allow farmers to maintain high yields, while using fewer amounts of chemical fertilizer and water. While critics point out the creation of super pests and weeds, which are highly possible, these dangers can be prevented and are manageable.
5. They can help reduce world hunger.-With genetic modification, even the smallest subsistence farmers will be able to produce larger, more reliable and profitable crops for the market, as well as able to grow more food for their families. This means that we will be able to directly address the problem of hunger and malnutrition among poor consumers, especially in

the developing countries. In the long run, genetically modified crops are considered to be a solution to ensuring that food production can keep pace with the growing population, which is expected to double 12 billion by 2050. The rate of growth of food production has declined in the past decade, with the latest round of increase driven by Green Revolution methods, such as using high-yielding hybrid seeds, extensive amounts of fertilizer and pesticides. However, these technologies do not have the capability to support food production growth for the coming decades without resorting to things that can severely damage the environment. Now, GM crops are believed to be able to support such growth.

List of Cons of Genetically Modified Crops

1. They are not entirely believed to be safe.-Truth is, there are also a number of independent studies of the health effects of genetically modified crops that do not have confidence in their safety. In an experiment in Scotland, for example, mice that were fed with modified potatoes containing lectin fared poorly and even suffered damage to their internal organs. While supporters have attacked the study, it entails the need for more research at the very least.
2. They contribute to the increased occurrence of allergic reactions. -Many studies show that consuming genetically modified crops has increased people's risk of acquiring allergies that are food-related, with symptoms including fatigue, rashes, headache and nausea.
3. They alter the way nature works.-Genetically engineered crops are believed to do wide-reaching damage to the environment. For example, insect-resistant plants might harm insects that are not their intended target, such as monarch butterflies. Also, target insects could develop resistance to these crops, requiring farmers to resort to using more aggressive measures to control the problem, such as extensive use of chemical sprays. As previously implied, the technology could also allow the transfer of genes from one crop to another, creating "super weeds", which will be immune to common control methods. Moreover, viral genes added to crops to confer resistance might be transferred to other viral pathogens, which can lead to new and more virulent virus strains.
4. They require labeling.-Food labeling can incite fear, causing the public not to accept food items made from genetically modified crops. As you can see, consumers might object to consuming of GM foods on the basis of

health, ethical and religious concerns. Lack of evidence that proves that GM products are unsafe must not be taken as proof that they are actually safe.

5. They only have little to no economic value.-While genetically modified crops might be resistant to herbicides and pesticides, as well as contain more nutrients, they actually contribute little to the farmers' profits. They take more time to mature and more effort to grow compared to natural crops.

Now, do you think that genetically modified crops are good for society or not? Do their benefits outdo the risks? By carefully weighing the pros and cons listed above, you can be a better judge.

Genetically modified brinjal

The genetically modified brinjal is a suite of transgenic brinjals (also known as an eggplant or aubergine) created by inserting a crystal protein gene (Cry1Ac) from the soil bacterium *Bacillus thuringiensis* into the genome of various brinjal cultivars. The insertion of the gene, along with other genetic elements such as promoters, terminators and an antibiotic resistance marker gene into the brinjal plant is accomplished using Agrobacterium-mediated genetic transformation. The Bt brinjal has been developed to give resistance against lepidopteron insects, in particular the Brinjal Fruit and Shoot Borer (Leucinodes orbonalis)(FSB). Mahyco, an Indian seed company based in Jalna, Maharashtra, has developed the Bt brinjal.

The genetically modified brinjal event is termed Event EE 1, and Mahyco have also applied for approval of two brinjal hybrids. The Event EE 1 was introgressed by plant breeding into various local varieties by University of Agricultural Sciences, Dharwad and Tamil Nadu Agricultural University, Coimbatore. Some of the cultivars of brinjal include: Malpur local, Manjari gota, Kudachi local, Udupi local, 112 GO, and Pabkavi local. It was approved for commercialization in India in 2009, but - after an apparent public outcry and rounds of debates in which representatives from Mahyco, the scientific community, and NGO's spoke on the topic - the then Indian Environment Minister, Jairam Ramesh, facilitated a moratorium on its release until further, unspecified, tests were conducted. Bt brinjal was approved for commercial release in Bangladesh in 2013.

Development

Mahyco licensed and used the cry1Ac gene obtained from Monsanto and two supporting genes (nptII and aad). The cry1Ac gene is under the transcriptional control of an enhanced cauliflower mosaic virus 35S (CaMV35S) promoter, which ensures the gene is expressed in all the brinjal's tissue throughout its complete life cycle. NptII and aad are selectable marker genes, nptII is used to identify transgenic plants from non-transgenic, and aad is used to identify the transformed bacteria used during the development of the construct. Aad contains a bacterial promoter and is not expressed in the Bt brinjal. The completed construct was inserted into young cotyledons from the brinjal plants using an Agrobacterium-mediated technique. Agrobacterium naturally inserts DNA into plants from its Ti plasmid, and scientists use this to insert genes of interest into various plants. The transformed plants were regenerated and analyzed for the presence of the gene through Southern blotting. The plants' progeny were also analyzed to identify lines segregating in a Mendelian fashion.

Effective against pests

When fruit and shoot borer larvae feed on Bt brinjal plants, they ingest the Bt protein Cry1Ac along with plant tissue. In the insect gut, which is alkaline with a pH >9.5, the protein is solubilized and activated by gut proteases. The Bt protein binds to specific receptor proteins present in the insect membrane, resulting in pore formation in the membranes. This leads to disruption of digestive processes, paralysis, and subsequent death of the fruit and shoot borer larvae.

Attempted commercialization in India

The first agreement to develop Bt Brinjal was signed in 2005 between India's leading seed company, Maharashtra Hybrid Seeds Company, better known as Mahyco, and two agricultural universities - University of Agricultural Sciences, Dharwad (UAS) and Tamil Nadu Agricultural University (TNAU) in Coimbatore.

An expert committee (EC-I) was set up in 2006 to examine the biosafety data presented by Mahyco. They concluded that while the current data demonstrated that Bt brinjal was safe and equivalent to its non-Bt counterpart, more studies were required to re-affirm the findings and further trials were needed to ascertain the benefits from Bt brinjal with respect to existing methods for pest management and pesticide reduction. They recommended that large scale trials be allowed to go ahead. In 2009, a second expert committee (EC-II) examined the data from these trials. They concluded that adequate safety tests had been performed, stating that "the benefits of Bt brinjal event EE-I developed by M/s Mahyco far outweigh the perceived and projected risks", and advised the Genetic Engineering Appraisal Committee (GEAC) to recommend commercialization of the Bt brinjal.

The GEAC cleared Bt brinjal for commercialization on 14 October 2009. Following concerns raised by some scientists, farmers and anti-GMO activists, the government of India officially announced on 9 February 2010 that it needed more time before releasing Bt brinjal, with Indian Environment Minister Jairam Ramesh saying that there is no overriding urgency to introduce Bt brinjal in India. On 17 February 2010, Jairam Ramesh reiterated that the Centre had only imposed a moratorium on the release of transgenic brinjal hybrid, and not a permanent ban, saying that "until we arrive at a political, scientific and societal consensus, this moratorium will remain". Companies with any seeds of Bt brinjal will have to register the details with the government, and the National Bureau of Plant Genetic Resources (NBPGR) was made responsible for storage of all the Bt brinjal seeds in India. Independent testing labs are currently being set up.

An irregularity was also brought to the notice of the Karnataka Biodiversity Board by Environment Support Group, a charitable trust in Bengaluru, in February 2010. It found that agencies accessed at least 10 brinjal varieties from Karnataka and Tamil Nadu without seeking prior consent of the National Biodiversity Authority and state biodiversity boards. Mahyco became India's first commercial entity to be accused of bio-piracy, or misappropriation, of local germplasm. In October 2013, the Indian High Court was pursuing criminal proceedings against senior officials of Mahyco-Monsanto.

Commercial Cultivation in Bangladesh

On 30 October 2013 with approvals from the ministries of Environment and Forests (MoEF) and Agriculture (MoA), the Bangladesh Agricultural Research Institute (BARI) received permission to release four varieties of Bt brinjal in time for the 2013–2014 growing season: Bt Uttara, Bt Kajla, Bt Nayantara, and Bt ISD006. The Bt varieties underwent seven years of field and greenhouse trials in various environmental and geographic locations in Bangladesh Saplings were distributed to 20 farmers in January 2014 The Guardian spoke to 19 of 20 farmers growing the Bt brinjal crop in 2014 and established that it has so far had mixed results. In September 2016 the Indian Business Standard quoted a Director of the Bangladesh Department of Environment who said that production results had been very good from the 200-300 farmers who had grown the bt brinjal since 2013 but it was too early to judge whether there had been any contamination of wild brinjal.

Controversy

Many controversies surround the development and release of genetically modified foods, ranging from human safety and environmental impacts to ethical concerns such as corporate control of the food supply and intellectual property rights. The brinjal is an important food crop for India, and the potential commercialization of a genetically modified variety has drawn support and criticism. Although it is a major food crop in India, brinjal production is relatively low with fruit and shoot borer infestation a major constraint to yield. Proponents of the technology believe the Bt brinjal will have positive effects for the Indian economy and the health of the farmers. Field trials conducted on research-managed farms carried out by Mahyco and the Indian Council of Agricultural Research suggested a 42% pesticide reduction and a doubling of the yield was possible. The economic gain for consumers, developers and farmers was estimated to potentially be US$108 million per year with an additional $3–4 million saved due to health benefits associated with decreased pesticide use.

A French scientist notable for his anti-GM perspective,Gilles-Eric Seralini, raised concerns about some of the differences between feeding trials using the genetically modified and unmodified brinjal, and criticized some of the testing protocols. The EC-II responded to the concerns raised by Seralini and other scientists in their report, New Zealand epidemiologist Lou Gallagher also criticised the feeding trials saying that the raw data indicated toxic effects were associated with the rats fed Bt Brinjal. Concerns have also been raised about a possible conflict of interest, with some of the scientists appointed to the GEAC being involved in developing their own GM products, that the decision by the EC-II was not unanimous, and about the reliability of safety data originating from Mahcyo run trials.

The imposed moratorium has been criticized by some scientists as not being based on any compelling scientific evidence and potentially setting Indian biotechnology back decades. Others feel the critical issue is not the safety of the GM technology, but its corporatization and there are claims that India's crop protection industry was a major player in preventing the commercialization of the Bt brinjal. India's National Biodiversity Authority is probing the crop scientists involved in developing the Bt brinjal for allegedly violating India's Biological Diversity Act, 2002 by using local cultivars and foreign technology without their permission. The Parliamentary Committee on Agriculture on 9 August 2012 asked the Government to stop all field trials and sought a ban on GM food crops like Bt brinjal. It also sought a "thorough probe" as to how permission was given to commercialise Bt brinjal seed when all evaluation tests were not carried out. The report of the Committee was tabled a day after Maharashtra Government cancelled Mahyco's licence to sell its Bt cotton seeds.

Genetically modified food controversies

Genetically modified food controversies are disputes over the use of foods and other goods derived from genetically modified crops instead of conventional crops, and other uses of genetic engineering in food production. The disputes involve consumers, farmers, biotechnology companies, governmental regulators, non-governmental organizations, and scientists. The key areas of controversy related to genetically modified food (GM food or GMO food) are whether such food should be labeled, the role of government regulators, the objectivity of scientific research and publication, the effect of genetically modified crops on health and the environment, the effect on pesticide resistance, the impact of such crops for farmers, and the role of the crops in feeding the world population. In addition, products derived from GMO organisms play a role in the production of ethanol fuels and pharmaceuticals.

Specific concerns include mixing of genetically modified and non-genetically modified products in the food supply, effects of GMOs on the environment, the rigor of the regulatory process, and consolidation of control of the food supply in companies that make and sell GMOs.Advocacy groups such as the Center for Food Safety, Organic Consumers Association, Union of Concerned Scientists, and Greenpeace say risks have not been adequately identified and managed, and they have questioned the objectivity of regulatory authorities.

The safety assessment of genetically engineered food products by regulatory bodies starts with an evaluation of whether or not the food is substantially equivalent to non-genetically engineered counterparts that are already deemed fit for human consumption. No reports of ill effects have been documented in the human population from genetically modified food. There is a scientific consensus that currently available food derived from GM crops poses no greater risk to human health than conventional food, but that each GM food needs to be tested on a case-by-case basis before introduction. Nonetheless, members of the public are much less likely than scientists to perceive GM foods as safe. The legal and regulatory status of GM foods varies by country, with some nations banning or restricting them, and others permitting them with widely differing degrees of regulation.

Public perception

Consumer concerns about food quality first became prominent long before the advent of GM foods in the 1990s. Upton Sinclair's novel The Jungle led to the 1906 Pure Food and Drug Act, the first major US legislation on the subject. This began an enduring concern over the purity and later "naturalness" of food that evolved from a single focus on sanitation to include others on added ingredients such as preservatives, flavors and sweeteners, residues such as pesticides, the rise of organic food as a category and, finally, concerns over GM food. Some consumers, including many in the US, came to see GM food as "unnatural", with various negative associations and fears (a reverse halo effect).

Specific perceptions include a view of genetic engineering as meddling with naturally evolved biological processes, and one that science has limitations on its comprehension of potential negative ramifications. An opposing perception is that genetic engineering is itself an evolution of traditional selective breeding, and that the weight of current evidence suggests current GM foods are identical to conventional foods in nutritional value and effects on health.

Surveys indicate widespread concern among consumers that eating genetically modified food is harmful, that biotechnology is risky, that more information is needed and that consumers need control over whether to take such risks. A diffuse sense that social and technological change is accelerating, and that people cannot affect this context of change, becomes focused when such changes affect food. Leaders in driving public perception of the harms of such food in the media include Jeffrey M. Smith, Dr. Oz, Oprah, and Bill Maher; organizations include Organic Consumers Association, Greenpeace (especially with regard to Golden rice) and Union of Concerned Scientists. In the United States support or opposition or skepticism about GMO food is not divided by traditional partisan (liberal/ conservative) lines, but young adults are more likely to have negative opinions on genetically modified food than older adults. Religious groups have raised concerns over whether genetically modified food will remain kosher or halal. In 2001, no such foods had been designated as unacceptable by Orthodox rabbis or Muslim leaders. However, some Jewish groups dispute this designation. Food writer Michael Pollan does not oppose eating genetically modified foods, but supports mandatory labeling of GM foods and has criticized the pesticide-heavy monoculture farming enabled by certain GM crops, such as glycosphate-tolerant ("Roundup-ready") corn and soybeans. He has also expressed concerns about biotechnology companies holding the intellectual property of the foods people depend on, and about the effects of the growing corporatization of large-scale agriculture. To address these problems, Pollan has brought up the idea of open sourcing GM foods. The idea has since been adopted to varying degrees by companies like Syngenta, and is being promoted by organizations such as the New America Foundation. Some organizations, like The BioBricks Foundation, have already worked out open-source licenses that could prove useful in this endeavour.

Public Relations Campaigns and Protests

In May 2012, a group called "Take the Flour Back" led by Gerald Miles protested plans by a group from Rothamsted Experimental Station, based in Harpenden, Hertfordshire, England, to conduct an experimental trial wheat genetically modified to repel aphids. The researchers, led by John Pickett, wrote a letter to the group in early May 2012, asking them to call off their protest, aimed for 27 May 2012. Group member Lucy Harrap said that the group was concerned about spread of the crops into nature, and cited examples of outcomes in the United States and Canada. Rothamsted Research and Sense About Science ran question and answer sessions about such a potential.

The March Against Monsanto is an international grassroots movement and protest against Monsanto corporation, a producer of genetically modified organism (GMOs) and Roundup, a glyphosate-based herbicide. The movement was founded by Tami Canal in response to the failure of California Proposition 37, a ballot initiative which would have required labeling food products made from GMOs. Advocates support mandatory labeling laws for food made from GMOs . The initial march took place on May 25, 2013. The number of protesters who took part is uncertain; figures of "hundreds of thousands" and the organizers' estimate of "two million" were variously cited. Events took place in between 330 and 436 cities around the world, mostly in the United States. Many protests occurred in Southern California, and some participants carried signs expressing support for mandatory labeling of GMOs that read "Label GMOs, It's Our Right to Know", and "Real Food 4 Real People". Canal said that the movement would continue its "anti-GMO cause" beyond the initial event. Further marches occurred in October 2013 and in May 2014 and 2015. The protests were reported by news outlets including ABC News, the Associated Press,The Washington Post,The Los Angeles Times,USA Today, and CNN (in the United States), and Russia Today and The Guardian (outside the United States).

Monsanto said that it respected people's rights to express their opinion on the topic, but maintained that its seeds improved agriculture by helping farmers produce more from their land while conserving resources, such as water and energy. The company reiterated that genetically modified foods were safe and improved crop yields. Similar sentiments were expressed by the Hawaii Crop Improvement Association, of which Monsanto is a member. In July 2013, the agricultural biotechnology industry launched a GMO transparency initiative called GMO Answers to address consumers' questions about GM foods in the U.S. food supply. GMO Answers' resources included conventional and organic farmers, agribusiness experts, scientists, academics, medical doctors and nutritionists, and "company experts" from founding members of the Council for Biotechnology Information, which funds the initiative. Founding members include BASF, Bayer CropScience, Dow AgroSciences, DuPont, Monsanto Company and Syngenta. In October 2013, a group called The European Network of Scientists for Social and Environmental Responsibility (ENSSER), posted a statement claiming that there is no scientific consensus on the safety of GMOs, which was signed by about 200 scientists in various fields in its first week. On January 25, 2015, their statement was formally published as a whitepaper by Environmental Sciences Europe:

Response to anti-GMO sentiment

In 2017, two documentaries were released which countered the growing anti-GMO sentiment among the public. These included Food Evolution and Science Moms. Per the Science Moms director, the film "focuses on providing a science and evidence-based counter-narrative to the pseudoscience-based parenting narrative that has cropped up in recent years".

Conspiracy theories

There are various conspiracy theories related to the production and sale of genetically modified crops and genetically modified food that have been identified by some commentators such as Michael Shermer. Generally, these conspiracy theories posit that GMOs are being knowingly and maliciously introduced into the food supply either as a means to unduly enrich agribusinesses or as a means to poison or pacify the population.

A work seeking to explore risk perception over GMOs in Turkey identified a belief among the conservative political and religious figures who were opposed to GMOs that GMOs were "a conspiracy by Jewish Multinational Companies and Israel for world domination."Additionally, a Latvian study showed that a segment of the population believed that GMOs were part of a greater conspiracy theory to poison the population of the country.

LAWSUITS

Foundation on Economic Trends v. Heckler

In 1983, environmental groups and protesters delayed the field tests of the genetically modified ice-minus strain of P. syringae with legal challenges.

Alliance for Bio-Integrity v. Shalala

In this case, the plaintiff argued both for mandatory labeling on the basis of consumer demand, and that GMO foods should undergo the same testing requirements as food additives because they are "materially changed" and have potentially unidentified health risks. The plaintiff also alleged that the FDA did not follow the Administrative Procedures Act in formulating and disseminating its policy on GMO's. The federal district court rejected all of those arguments and found that the FDA's determination that GMO's are Generally Recognized as Safe was neither arbitrary nor capricious. The court gave deference to the FDA's process on all issues, leaving future plaintiffs little legal recourse to challenge the FDA's policy on GMO's.

Diamond v. Chakrabarty

The Diamond v. Chakrabarty case was on the question of whether GMOs can be patented. On 16 June 1980, the Supreme Court, in a 5–4 split decision, held that "A live, human-made micro-organism is patentable subject matter" under the meaning of U.S. patent law.

Bt maize

One of the first incidents occurred in 1999, when Nature published a paper on potential toxic effects of Bt maize on butterflies. The paper produced a public uproar and demonstrations, however by 2001 multiple follow-up studies had concluded that "the most common types of Bt maize pollen are not toxic to monarchlarvae in concentrations the insects would encounter in the fields" and that they had "brought that particular question to a close".

Concerned scientists began to patrol the scientific literature and react strongly, both publicly and privately, to discredit conclusions they view as flawed in order to prevent unjustified public outcry and regulatory action. A 2013 Scientific American article noted that a "tiny minority" of biologists have published concerns about GM food, and said that scientists who support the use of GMOs in food production are often overly dismissive of them.

Restrictive end-user agreements

Prior to 2010, scientists wishing to conduct research on commercial GM plants or seeds were unable to do so, because of restrictive end-user agreements. Cornell University's Elson Shields was the spokesperson for one group of scientists who opposed such restrictions. The group submitted a statement to the United States Environmental Protection Agency (EPA) in 2009 protesting that "as a result of restrictive access, no truly independent research can be legally conducted on many critical questions regarding the technology".

A 2009 Scientific American editorial quoted a scientist who said that several studies that were initially approved by seed companies were blocked from publication when they returned "unflattering" results. While favoring protection of intellectual property rights, the editors called for the restrictions to be lifted and for the EPA to require, as a condition of approval, that independent researchers have unfettered access to genetically modified products for research. In December 2009, the American Seed Trade Association agreed to "allow public researchers greater freedom to study the effects of GM food crops". The companies signed blanket agreements permitting such research. This agreement left many scientists optimistic about the future; other scientists still express concern as to whether this agreement has the ability to "alter what has been a research environment rife with obstruction and suspicion". Monsanto previously had research agreements (i.e., Academic Research Licenses) with approximately 100 universities that allowed for university scientists to conduct research on their GM products with no oversight.

Reviews

A 2011 analysis by Diels et al., reviewed 94 peer-reviewed studies pertaining to GMO safety to assess whether conflicts of interest correlated with outcomes that cast GMOs in a favorable light. They found that financial conflict of interest was not associated with study outcome ($p = 0.631$) while author affiliation to industry (i.e., a professional conflict of interest) was strongly associated with study outcome ($p < 0.001$).Of the 94 studies that were analyzed, 52% did not declare funding. 10% of the studies were categorized as "undetermined" with regard to professional conflict of interest. Of the 43 studies with financial or professional conflicts of interest, 28 studies were compositional studies. According to Marc Brazeau, an association between professional conflict of interest and positive study outcomes can be skewed because companies typically contract with independent researchers to perform follow-up studies only after in-house research uncovers favorable results. In-house research that uncovers negative or unfavorable results for a novel GMO is generally not further pursued. A 2013 review, of 1,783 papers on genetically modified crops and food published between 2002 and 2012 found no plausible evidence of dangers from the use of then marketed GM crops. Biofortified, an independent nonprofit organization devoted to providing factual information and fostering discussion about agriculture, especially plant genetics and genetic engineering, planned to add the studies found by the Italian group to its database of studies about GM crops, GENERA.

In a 2014 review, Zdziarski et al. examined 21 published studies of the histopathology of GI tracts of rats that were fed diets derived from GM crops, and identified some systemic flaws in this area of the scientific literature. Most studies were performed years after the approval of the crop for human consumption. Papers were often imprecise in their descriptions of the histological results and the selection of study endpoints, and lacked necessary details about methods and results. The authors called for the development of better study guidelines for determining the long-term safety of eating GM foods.

A 2016 study by the US National Academies of Sciences, Engineering, and Medicine concluded that GM foods are safe for human consumption and they could find no conclusive evidence that they harm the environment nor wildlife. They analysed over 1.000 studies over the previous 30 years that GM crops have been available, reviewed 700 written presentations submitted by interested bodies and heard 80 witnesses. They concluded that GM crops had given farmers economic advantages but found no evidence that GM crops had increased yields. They also noted that weed resistance to GM crops could cause major agricultural problems but this could be addressed by better farming procedures.

ALLEGED DATA MANIPULATION

A University of Naples investigation suggested that images in eight papers on animals were intentionally altered and/or misused. The leader of the research group, Federico Infascelli, rejected the claim. The research concluded that mother goats fed GM soybean meal secreted fragments of the foreign gene in their milk. In December 2015 one of the papers was retracted for "self-plagiarism", although the journal noted that the results remained valid.A second paper was retracted in March 2016 after The University of Naples concluded that "multiple heterogeneities were likely attributable to digital manipulation, raising serious doubts on the reliability of the findings".

Substantial equivalence

Most conventional agricultural products are the products of genetic manipulation via traditional cross-breeding and hybridization. Governments manage the marketing and release of GM foods on a case-by-case basis. Countries differ in their risk assessments and regulations. Marked differences distinguish the US from Europe. Crops not intended as foods are generally not reviewed for food safety. GM foods are not tested in humans before marketing because they are not a single chemical, nor are they intended to be ingested using specific doses and intervals, which complicate clinical study design. Regulators examine the genetic modification, related protein products and any changes that those proteins make to the food.

Regulators check that GM foods are "substantially equivalent" to their conventional counterparts, to detect any negative unintended consequences. New protein(s) that differ from conventional food proteins or anomalies that arise in the substantial equivalence comparison require further toxicological analysis.

"The World Health Organization, the American Medical Association, the U.S. National Academy of Sciences, the British Royal Society, and every other respected organization that has examined the evidence has come to the same conclusion: consuming foods containing ingredients derived from GM crops is no riskier than consuming the same foods containing ingredients from crop plants modified by conventional plant improvement techniques."

–American Association for the Advancement of Science

In 1999, Andrew Chesson of the Rowett Research Institute warned that substantial equivalence testing "could be flawed in some cases" and that current safety tests could allow harmful substances to enter the human food supply. The same year Millstone, Brunner and Mayer argued that the standard was a pseudo-scientific product of politics and lobbying that was created to reassure consumers and aid biotechnology companies to reduce the time and cost of safety testing. They suggested that GM foods have extensive biological, toxicological and immunological tests and that substantial equivalence should be abandoned. This commentary was criticized for misrepresenting history, for distorting existing data and poor logic. Kuiper claimed that it oversimplified safety assessments and that equivalence testing involves more than chemical tests, possibly including toxicity testing. Keler and Lappe supported Congressional legislation to replace the substantial equivalence standard with safety studies.In a 2016 review, Domingo criticized the use of the "substantial equivalence" concept as a measure of the safety of GM crops.

Kuiper examined this process further in 2002, finding that substantial equivalence does not measure absolute risks, but instead identifies differences between new and existing products. He claimed that characterizing differences is properly a starting point for a safety assessment and "the concept of substantial equivalence is an adequate tool in order to identify safety issues related to genetically modified products that have a traditional counterpart". Kuiper noted practical difficulties in applying this standard, including the fact that traditional foods contain many toxic or carcinogenic chemicals and that existing diets were never proven to be safe. This lack of knowledge re conventional food means that modified foods may differ in anti-nutrients and natural toxins that have never been identified in the original plant, possibly allowing harmful changes to be missed. In turn, positive modifications may also be missed. For example, corn damaged by insects often contains high levels of fumonisins, carcinogenic toxins made by fungi that travel on insects' backs and that grow in the wounds of damaged corn. Studies show that most Bt corn has lower levels of fumonisins than conventional insect-damaged corn. Workshops and consultations organized by the OECD, WHO, and FAO have worked to acquire data and develop better understanding of conventional foods, for use in assessing GM foods.

A survey of publications comparing the intrinsic qualities of modified and conventional crop lines (examining genomes, proteomes and metabolomes) concluded that GM crops had less impact on gene expression or on protein and metabolite levels than the variability generated by conventional breeding.

In a 2013 review, Herman (Dow AgroSciences) and Price (FDA, retired) argued that transgenesis is less disruptive than traditional breeding techniques because the latter routinely involve more changes (mutations, deletions, insertions and rearrangements) than the relatively limited changes (often single gene) in genetic engineering. The FDA found that all of the 148 transgenic events that they evaluated to be substantially equivalent to their conventional counterparts, as have Japanese regulators for 189 submissions including combined-trait products. This equivalence was confirmed by more than 80 peer-reviewed publications. Hence, the authors argue, compositional equivalence studies uniquely required for GM food crops may no longer be justified on the basis of scientific uncertainty.

Allergenicity

A well-known risk of genetic modification is the introduction of an allergen. Allergen testing is routine for products intended for food, and passing those tests is part of the regulatory requirements. Organizations such as the European Green Party and Greenpeace emphasize this risk.A 2005 review of the results from allergen testing stated that "no biotech proteins in foods have been documented to cause allergic reactions". Regulatory authorities require that new modified foods be tested for allergenicity before they are marketed.

GMO proponents note that because of the safety testing requirements, the risk of introducing a plant variety with a new allergen or toxin is much smaller than from traditional breeding processes, which do not require such tests. Genetic engineering can have less impact on the expression of genomes or on protein and metabolite levels than conventional breeding or (non-directed) plant mutagenesis. Toxicologists note that "conventional food is not risk-free; allergies occur with many known and even new conventional foods. For example, the kiwi fruit was introduced into the U.S. and the European markets in the 1960s with no known human allergies; however, today there are people allergic to this fruit."

Genetic modification can also be used to remove allergens from foods, potentially reducing the risk of food allergies.A hypo-allergenic strain of soybean was tested in 2003 and shown to lack the major allergen that is found in the beans. A similar approach has been tried in ryegrass, which produces pollen that is a major cause of hay fever: here a fertile GM grass was produced that lacked the main pollen allergen, demonstrating that hypoallergenic grass is also possible.

The development of genetically modified products found to cause allergic reactions has been halted by the companies developing them before they were brought to market. In the early 1990s, Pioneer Hi-Bred attempted to improve the nutrition content of soybeans intended for animal feed by adding a gene from the Brazil nut. Because they knew that people have allergies to nuts, Pioneer ran in vitro and skin prick allergy tests. The tests showed that the transgenic soy was allergenic. Pioneer Hi-Bred therefore discontinued further development. In 2005, a pest-resistant field pea developed by the Australian Commonwealth Scientific and Industrial Research Organisation for use as a pasture crop was shown to cause an allergic reaction in mice. Work on this variety was immediately halted. These cases have been used as evidence that genetic modification can produce unexpected and dangerous changes in foods, and as evidence that safety tests effectively protect the food supply.

During the Starlink corn recalls in 2000, a variety of GM maize containing the Bacillus thuringiensis (Bt) protein Cry9C, was found contaminating corn products in U.S. supermarkets and restaurants. It was also found in Japan and South Korea. Starlink corn had only been approved for animal feed as the Cry9C protein lasts longer in the digestive system than other Bt proteins raising concerns about its potential allergenicity.In 2000, Taco Bell-branded taco shells sold in supermarkets were found to contain Starlink, resulting in a recall of those products, and eventually led to the recall of over 300 products. Sales of StarLink seed were discontinued and the registration for the Starlink varieties was voluntarily withdrawn by Aventis in October 2000. Aid sent by the United Nations and the United States to Central African nations was also found to be contaminated with StarLink corn and the aid was rejected. The U.S. corn supply has been monitored for Starlink Bt proteins since 2001 and no positive samples have been found since 2004. In response, GeneWatch UK and Greenpeace set up the GM Contamination Register in 2005. During the recall, the United States Centers for Disease Control evaluated reports of allergic reactions to StarLink corn, and determined that no allergic reactions to the corn had occurred.

Horizontal gene Transfer

Horizontal gene transfer is the movement of genes from one organism to another in a manner other than reproduction. The risk of horizontal gene transfer between GMO plants and animals is very low and in most cases is expected to be lower than background rates. Two studies on the possible effects of feeding animals with genetically modified food found no residues of recombinant DNA or novel proteins in any organ or tissue samples. Studies found DNA from the M13 virus, Green fluorescent protein and RuBisCO genes in the blood and tissue of animals, and in 2012, a paper suggested that a specific microRNA from rice could be found at very low quantities in human and animal serum. Other studies however, found no or negligible transfer of plant microRNAs into the blood of humans or any of three model organisms. Another concern is that the antibiotic resistance gene commonly used as a genetic marker in transgenic crops could be transferred to harmful bacteria, creating resistant superbugs. A 2004 study involving human volunteers examined whether the transgene from modified soy would transfer to bacteria that live in the human gut. As of 2012 it was the only human feeding study to have been conducted with GM food. The transgene was detected in three volunteers from a group of seven who had previously had their large intestines removed for medical reasons. As this gene transfer did not increase after the consumption of the modified soy, the researchers concluded that gene transfer did not occur. In volunteers with intact digestive tracts, the transgene did not survive. The antibiotic resistance genes used in genetic engineering are naturally found in many pathogens and antibiotics these genes confer resistance to are not widely prescribed.

CONTROVERSIAL STUDIES

Pusztai Affair

Árpád Pusztai published the first peer-reviewed paper to find negative effects from GM food consumption in 1999. Pusztai fed rats potatoes transformed with the Galanthus nivalisagglutinin (GNA) gene from the Galanthus (snowdrop) plant, allowing the tuber to synthesise the GNA lectin protein.While some companies were considering growing GM crops expressing lectin, GNA was an unlikely candidate.Lectin is toxic, especially to gut epithelia. Pusztai reported significant differences in the thickness of the gut epithelium, but no differences in growth or immune system function.

On June 22, 1998, an interview on Granada Television's current affairs programme World in Action, Pusztai said that rats fed on the potatoes had stunted growth and a repressed immune system.Amedia frenzy resulted. Pusztai was suspended from the Rowett Institute. Misconduct procedures were used to seize his data and ban him from speaking publicly. The Rowett Institute and the Royal Society reviewed his work and concluded that the data did not support his conclusions. The work was criticized on the grounds that the unmodified potatoes were not a fair control diet and that any rat fed only potatoes would suffer from protein deficiency. Pusztai responded by stating that all diets had the same protein and energy content and that the food intake of all rats was the same.

Bt corn

A 2011 study, was the first to evaluate the correlation between maternal and fetal exposure to Bt toxin produced in GM maize and to determine exposure levels of the pesticides and their metabolites. It reported the presence of pesticides associated with the modified foods in women and in pregnant women's fetuses. The paper and related media reports were criticized for overstating the results. Food Standards Australia New Zealand (FSANZ) posted a direct response, saying that the suitability of the ELISA method for detecting the Cry1Ab protein was not validated and that no evidence showed that GM food was the protein's source. The organization also suggested that even had the protein been detected its source was more likely conventional or organic food.

Séralini affair

In 2007, 2009, and 2011, Gilles-Éric Séralini published re-analysis studies that used data from Monsanto rat-feeding experiments for three modified maize varieties (insect-resistant MON 863 and MON 810 and glyphosate-resistant NK603). He concluded that the data showed liver, kidney and heart damage. The European Food Safety Authority (EFSA) then concluded that the differences were all within the normal range. EFSA also stated that Séralini's statistics were faulty. EFSA's conclusions were supported by FSANZ, a panel of expert toxicologists, and the French High Council of Biotechnologies Scientific Committee (HCB).

In 2012, Séralini's lab published a paper that considered the long-term effects of feeding rats various levels of GM glyphosate-resistant maize, conventional glyphosate-treated maize, and a mixture of the two strains.The paper concluded that rats fed the modified maize had severe health problems, including liver and kidney damage and large tumors.The study provoked widespread criticism. Séralini held a press conference just before the paper was released in which he announced the release of a book and a movie. He allowed reporters to have access to the paper before his press conference only if they signed a confidentiality agreement under which they could not report other scientists' responses to the paper.The press conference resulted in media coverage emphasizing a connection between GMOs, glyphosate, and cancer. Séralini's publicity stunt yielded criticism from other scientists for prohibiting critical commentary. Criticisms included insufficient statistical power and that Séralini's Sprague-Dawley rats were inappropriate for a lifetime study (as opposed to a shorter toxicity study) because of their tendency to develop cancer (one study found that more than 80% normally got cancer). The Organisation for Economic Co-operation and Development guidelines recommended using 65 rats per experiment instead of the 10 in Séralini's. Other criticisms included the lack of data regarding food amounts and specimen growth rates, the lack of a dose–response relationship (females fed three times the standard dose showed a decreased number of tumours) and no identified mechanism for the tumour increases. Six French national academies of science issued an unprecedented joint statement condemning the study and the journal that published it.Food and Chemical Toxicology published many critical letters, with only a few expressing support. National food safety and regulatory agencies also reviewed the paper and dismissed it. In March 2013, Séralini responded to these criticisms in the same journal that originally published his study, and a few scientists supported his work. In November 2013, the editors of Food and Chemical Toxicology retracted the paper. The retraction was met with protests from Séralini and his supporters. In 2014, the study was republished by a different journal, Environmental Sciences Europe, in an expanded form, including the raw data that Séralini had originally refused to reveal.

NUTRITIONAL QUALITY

Some plants are specifically genetically modified to be healthier than conventional crops. Golden rice was created to combat vitamin A deficiency by synthesizing beta carotene (which conventional rice does not).

Detoxification

One variety of cottonseed has been genetically modified to remove the toxin gossypol, so that it would be safe for humans to eat.

Environment

Genetically modified crops are planted in fields much like regular crops. There they interact directly with organisms that feed on the crops and indirectly with other organisms in the food chain. The pollen from the plants is distributed in the environment like that of any other crop. This distribution has led to concerns over the effects of GM crops on the environment. Potential effects include gene flow, pesticide resistance and greenhouse gas emissions.

NON-TARGET ORGANISMS

A major use of GM crops is in insect control through the expression of the cry (crystal delta-endotoxins) and Vip (vegetative insecticidal proteins) genes from Bacillus thuringiensis (Bt). Such toxins could affect other insects in addition to targeted pests such as the European corn borer. Bt proteins have been used as organic sprays for insect control in France since 1938 and the US since 1958, with no reported ill effects.Cry proteins selectively target Lepidopterans (moths and butterflies). As a toxic mechanism, cry proteins bind to specific receptors on the membranes of mid-gut (epithelial) cells, resulting in their rupture. Any organism that lacks the appropriate receptors in its gut is unaffected by the cry protein, and therefore is not affected by Bt. Regulatory agencies assess the potential for transgenic plants to affect non-target organisms before approving their commercial release.

In 1999, a paper stated that, in a laboratory environment, pollen from Bt maize dusted onto milkweed could harm the monarch butterfly. A collaborative research exercise over the following two years by several groups of scientists in the US and Canada studied the effects of Bt pollen in both the field and the laboratory. The study resulted in a risk assessment concluding that any risk posed to butterfly populations was negligible. A 2002 review of the scientific literature concluded that "the commercial large-scale cultivation of current Bt–maize hybrids did not pose a significant risk to the monarch population" and noted that despite large-scale planting of genetically modified crops, the butterfly's population was increasing. However, the herbicide glyphosate used to grow GMOs kills milkweed, the only food source of monarch butterflies, and by 2015 about 90% of the U.S. population has declined. Lövei et al. analyzed laboratory settings and found that Bt toxins could affect non-target organisms, generally closely related to the intended targets.Typically, exposure occurs through the consumption of plant parts, such as pollen or plant debris, or through Bt ingestion by predators. A group of academic scientists criticized the analysis, writing: "We are deeply concerned about the inappropriate methods used in their paper, the lack of ecological context, and the authors' advocacy of how laboratory studies on non-target arthropods should be conducted and interpreted".

Biodiversity

Crop genetic diversity might decrease due to the development of superior GM strains that crowd others out of the market. Indirect effects might affect other organisms. To the extent that agrochemicals impact biodiversity, modifications that increase their use, either because successful strains require them or because the accompanying development of resistance will require increased amounts of chemicals to offset increased resistance in target organisms.

Studies comparing the genetic diversity of cotton found that in the US diversity has either increased or stayed the same, while in India it has declined. This difference was attributed to the larger number of modified varieties in the US compared to India.A review of the effects of Bt crops on soil ecosystems found that in general they "appear to have no consistent, significant, and long-term effects on the microbiota and their activities in soil".

The diversity and number of weed populations has been shown to decrease in farm-scale trials in the United Kingdom and in Denmark when comparing herbicide-resistant crops to their conventional counterparts. The UK trial suggested that the diversity of birds could be adversely affected by the decrease in weed seeds available for foraging. Published farm data involved in the trials showed that seed-eating birds were more abundant on conventional maize after the application of the herbicide, but that there were no significant differences in any other crop or prior to herbicide treatment. A 2012 study found a correlation between the reduction of milkweed in farms that grew glyphosate-resistant crops and the decline in adult monarch butterfly populations in Mexico.The New York Times reported that the study "raises the somewhat radical notion that perhaps weeds on farms should be protected. A 2005 study, designed to "simulate the impact of a direct overspray on a wetland" with four different agrochemicals (carbaryl (Sevin), malathion, 2,4-dichlorophenoxyacetic acid, and glyphosate in a Roundup formulation) by creating artificial ecosystems in tanks and then applying "each chemical at the manufacturer's maximum recommended application rates" found that "species richness was reduced by 15% with Sevin, 30% with malathion, and 22% with Roundup, whereas 2,4-D had no effect".The study has been used by environmental groups to argue that use of agrochemicals causes unintended harm to the environment and to biodiversity.

Secondary pests

Several studies documented surges in secondary pests within a few years of adoption of Bt cotton. In China, the main problem has been with mirids, which have in some cases "completely eroded all benefits from Bt cotton cultivation".A 2009 study in China concluded that the increase in secondary pests depended on local temperature and rainfall conditions and occurred in half the villages studied. The increase in insecticide use for the control of these secondary insects was far smaller than the reduction in total insecticide use due to Bt cotton adoption. A 2011 study based on a survey of 1,000 randomly selected farm households in five provinces in China found that the reduction in pesticide use in Bt cotton cultivars was significantly lower than that reported in research elsewhere: The finding was consistent with a hypothesis that more pesticide sprayings are needed over time to control emerging secondary pests, such as aphids, spider mites, and lygus bugs. Similar problems have been reported in India, with mealy bugs and aphids.

Gene flow

Genes from a GMO may pass to another organism just like an endogenous gene. The process is known as outcrossing and can occur in any new open-pollinated crop variety. Introduced traits potentially can cross into neighboring plants of the same or closely related species through three different types of gene flow: crop-to-crop, crop-to-weedy, and crop-to-wild. In crop-to-crop, genetic information from a genetically modified crop is transferred to a non-genetically modified crop. Crop-to-weedy transfer refers to the transfer of genetically modified material to a weed, and crop-to-wild indicates transfer from a genetically modified crop to a wild, undomesticated plant and/or crop. There are concerns that the spread of genes from modified organisms to unmodified relatives could produce species of weeds resistant to herbicides that could contaminate nearby non-genetically modified crops, or could disrupt the ecosystem, This is primarily a concern if the transgenic organism has a significant survival capacity and can increase in frequency and persist in natural populations.This process, whereby genes are transferred from GMOs to wild relatives, is different from the development of so-called "superweeds" or "superbugs" that develop resistance to pesticides under natural selection.

In most countries environmental studies are required before approval of a GMO for commercial purposes, and a monitoring plan must be presented to identify unanticipated gene flow effects.

In 2004, Chilcutt and Tabashnik found Bt protein in kernels of in a refuge (a conventional crop planted to harbor pests that might otherwise become resistant a pesticide associated with the GMO) implying that gene flow had occurred.

In 2005, scientists at the UK Centre for Ecology and Hydrology reported the first evidence of horizontal gene transfer of pesticide resistance to weeds, in a few plants from a single season; they found no evidence that any of the hybrids had survived in subsequent seasons.

In 2007, the U.S. Department of Agriculture fined Scotts Miracle-Gro $500,000 when modified DNA from GM creeping bentgrass, was found within relatives of the same genus (Agrostis) as well as in native grasses up to 21 km (13 mi) from the test sites, released when freshly cut, wind-blown grass.

In 2009, Mexico created a regulatory pathway for GM maize, but because Mexico is maize's center of diversity, concerns were raised about GM maize's effects on local strains. A 2001 report found Bt maize cross-breeding with conventional maize in Mexico.The data in this paper was later described as originating from an artifact and the publishing journal Nature stated that "the evidence available is not sufficient to justify the publication of the original paper", although it did not retract the paper. A subsequent large-scale study, in 2005, found no evidence of gene flow in Oaxaca. However, other authors claimed to have found evidence of such gene flow.

A 2010 study showed that about 83 percent of wild or weedy canola tested contained genetically modified herbicide resistance genes. According to the researchers, the lack of reports in the United States suggested that oversight and monitoring were inadequate.A 2010 report stated that the advent of glyphosate-resistant weeds could cause GM crops to lose their effectiveness unless farmers combined glyphosate with other weed-management strategies.

One way to avoid environmental contamination is genetic use restriction technology (GURT), also called "Terminator".This uncommercialized technology would allow the production of crops with sterile seeds, which would prevent the escape of GM traits. Groups concerned about food supplies had expressed concern that the technology would be used to limit access to fertile seeds. Another hypothetical technology known as "Traitor" or "T-GURT", would not render seeds sterile, but instead would require application of a chemical to GM crops to activate engineered traits. Groups such as Rural Advancement Foundation International raised concerns that further food safety and environmental testing needed to be done before T-GURT would be commercialized.

Escape of Modified Crops

The escape of genetically modified seed into neighboring fields, and the mixing of harvested products, is of concern to farmers who sell to countries that do not allow GMO imports. In 1999 scientists in Thailand claimed they had discovered unapproved glyphosate-resistant GM wheat in a grain shipment, even though it was only grown in test plots. No mechanism for the escape was identified. In 2000, AventisStarLink GM corn was found in US markets and restaurants. It became the subject of a recall that started when Taco Bell-branded taco shells sold in supermarkets were found to contain it. StarLink was then discontinued. Registration for Starlink varieties was voluntarily withdrawn by Aventis in October 2000.

American rice exports to Europe were interrupted in 2006 when the LibertyLink modification was found in commercial rice crops, although it had not been approved for release.An investigation by the USDA's Animal and Plant Health Inspection Service (APHIS) failed to determine the cause of the contamination.

In May 2013, unapproved glyphosate-resistant GM wheat (but that had been approved for human consumption)was discovered in a farm in Oregon in a field that had been planted with winter wheat. The strain was developed by Monsanto, and had been field-tested from 1998 to 2005. The discovery threatened US wheat exports which totaled $8.1 billion in 2012. Japan, South Korea and Taiwan temporarily suspended winter wheat purchases as a result of the discovery. As of August 30, 2013, while the source of the modified wheat remained unknown, Japan, South Korea and Taiwan had resumed placing orders.

COEXISTENCE WITH CONVENTIONAL CROPS

The US has no legislation governing the relationship among mixtures of farms that grow organic, conventional, and GM crops. The country relies on a "complex but relaxed" combination of three federal agencies (FDA, EPA, and USDA/APHIS) and states' common law tort systems to manage coexistence. The Secretary of Agriculture convened an Advisory Committee on Biotechnology and 21st Century Agriculture (AC21) to study coexistence and make recommendations about the issue. The members of AC21 included representatives of the biotechnology industry, the organic food industry, farming communities, the seed industry, food manufacturers, State governments, consumer and community development groups, the medical profession, and academic researchers. AC21 recommended that a study assess the potential for economic losses to US organic farmers; that any serious losses lead to a crop insurance program, an education program to ensure that organic farmers put appropriate contracts in place and that neighboring GMO farmers take appropriate containment measures. Overall the report supported a diverse agriculture system supporting diverse farming systems.

The EU implemented regulations specifically governing co-existence and traceability. Traceability has become commonplace in the food and feed supply chains of most countries, but GMO traceability is more challenging given strict legal thresholds for unwanted mixing. Since 2001, conventional and organic food and feedstuffs can contain up to 0.9% of authorised modified material without carrying a GMO label. (any trace of non-authorised modification is cause for a shipment to be rejected). Authorities require the ability to trace, detect and identify GMOs, and the several countries and interested parties created a non-governmental organization, Co-Extra, to develop such methods.

CHEMICAL USE

Pesticides

Pesticides destroy, repel or mitigate pests (an organism that attacks or competes with a crop). A 2014 meta-analysis covering 147 original studies of farm surveys and field trials, and 15 studies from the researchers conducting the study, concluded that adoption of GM technology had reduced chemical pesticide use by 37%, with the effect larger for insect-tolerant crops than herbicide-tolerant crops.Some doubt still remains on whether the reduced amounts of pesticides used actually invoke a lower negative environmental effect, since there is also a shift in the types of pesticides used, and different pesticides have different environmental effects. In August 2015, protests occurred in Hawaii over the possibility that birth defects were being caused by the heavy use of pesticides on new strains of GM crops being developed there. Hawaii uses 17 times the amount of pesticides per acre compared to the rest of the US.

Herbicides

The development of glyphosate-tolerant (Roundup Ready) plants changed the herbicide use profile away from more persistent, higher toxicity herbicides, such as atrazine, metribuzin and alachlor, and reduced the volume and harm of herbicide runoff.A study by Chuck Benbrook concluded that the spread of glyphosate-resistant weeds had increased US herbicide use. That study cited a 23% increase (.3 kilograms/hectare) for soybeans from 1996–2006, a 43% (.9 kg/ha) increase for cotton from 1996–2010 and a 16% (.5 kg/ha) decrease for corn from 1996–2010. However, this study came under scrutiny because Benbrook did not consider the fact that glyphosate is less toxic than other herbicides, thus net toxicity may decrease even as use increases. Graham Brookes accused Benbrook of subjective herbicide estimates because his data, provided by the National Agricultural Statistics Service, does not distinguish between genetically modified and non-genetically modified crops. Brookes had earlier published a study that found that the use of biotech crops had reduced the volume and environmental impact of herbicide and other pesticides, which contradicted Benbrook. Brookes stated that Benbrook had made "biased and inaccurate" assumptions.

Insecticides

A claimed environmental benefit of Bt-cotton and maize is reduced insecticide use. A PG Economics study concluded that global pesticide use was reduced by 286,000 tons in 2006, decreasing pesticidal environmental impact by 15%.A survey of small Indian farms between 2002 and 2008 concluded that Bt cotton adoption had led to higher yields and lower pesticide use. Another study concluded that insecticide use on cotton and corn during the years 1996 to 2005 fell by 35,600,000 kilograms (78,500,000 lb) of active ingredient, roughly equal to the annual amount applied in the European Union. A Bt cotton study in six northern Chinese provinces from 1990 to 2010 concluded that it halved the use of pesticides and doubled the level of ladybirds, lacewings and spiders and extended environmental benefits to neighbouring crops of maize, peanuts and soybeans.

Resistant insect pests

Resistance evolves naturally after a population has been subjected to selection pressure via repeated use of a single pesticide. In November 2009, Monsanto scientists found that the pink bollworm had become resistant to first generation Bt cotton in parts of Gujarat, India—that generation expresses one Bt gene, Cry1Ac. This was the first instance of Bt resistance confirmed by Monsanto. Similar resistance was later identified in Australia, China, Spain and the US.

One strategy to delay Bt-resistance is to plant pest refuges using conventional crops, thereby diluting any resistant genes. Another is to develop crops with multiple Bt genes that target different receptors within the insect.In 2012, a Florida field trial demonstrated that army worms were resistant to Dupont-Dow's GM corn. This resistance was discovered in Puerto Rico in 2006, prompting Dow and DuPont to stop selling the product there. The European corn borer, one of Bt's primary targets, is also capable of developing resistance.

Economy

GM food's economic value to farmers is one of its major benefits, including in developing nations. A 2010 study found that Bt corn provided economic benefits of $6.9 billion over the previous 14 years in five Midwestern states. The majority ($4.3 billion) accrued to farmers producing non-Bt corn. This was attributed to European corn borer populations reduced by exposure to Bt corn, leaving fewer to attack conventional corn nearby. Agriculture economists calculated that "world surplus [increased by] $240.3 million for 1996. Of this total, the largest share (59%) went to U.S. farmers. Seed company Monsanto received the next largest share (21%), followed by US consumers (9%), the rest of the world (6%), and the germplasm supplier, Delta and Pine Land Company (5%)." PG Economics comprehensive 2012 study concluded that GM crops increased farm incomes worldwide by $14 billion in 2010, with over half this total going to farmers in developing countries.

The main Bt crop grown by small farmers in developing countries is cotton. A 2006 review of Bt cotton findings by agricultural economists concluded, "the overall balance sheet, though promising, is mixed. Economic returns are highly variable over years, farm type, and geographical location". However, environmental activist Mark Lynas said that complete rejection of genetic engineering is "illogical and potentially harmful to the interests of poorer peoples and the environment".

In 2013, the European Academies Science Advisory Council (EASAC) asked the EU to allow the development of agricultural GM technologies to enable more sustainable agriculture, by employing fewer land, water and nutrient resources. EASAC also criticizes the EU's "timeconsuming and expensive regulatory framework" and said that the EU had fallen behind in the adoption of GM technologies.

Industrial agriculture

GM crops play a key role in intensive crop farming, which involves monoculture, use of herbicides and pesticides, use of equipment requiring large amounts of fuel and irrigation. Opponents such as Jonathan Latham of the Bioscience Resource Project and Vandana Shiva treat industrial agriculture and modified crops as closely related topics, and call for agriculture to adopt practices drastically reducing disruptions to the environment. Proponents of conventional agriculture point to its high yields, low prices, and wider choices and claim that technology is necessary to feed a growing world population.

Developing nations

Disagreements about developing nations include the claimed need for increased food supplies, and how to achieve such an increase. Some scientists suggest that a second Green Revolution including use of modified crops is needed to provide sufficient food. The potential for genetically modified food to help developing nations was recognised by the International Assessment of Agricultural Science and Technology for Development, but as of 2008 they had found no conclusive evidence of a solution. Skeptics such as John Avise claim that apparent shortages are caused by problems in food distribution and politics, rather than production. Other critics say that the world has so many people because the second green revolution adopted unsustainable agricultural practices that left the world with more mouths to feed than the planet can sustain. Pfeiffer claimed that even if technological farming could feed the current population, its dependence on fossil fuels, which in 2006 he incorrectly predicted would reach peak output in 2010, would lead to a catastrophic rise in energy and food prices.

Claimed deployment constraints to developing nations include the lack of easy access, equipment costs and intellectual property rights that hurt developing countries. The Consultative Group on International Agricultural Research (CGIAR), an aid and research organization, was praised by the World Bank for its efforts, but the bank recommended that they shift to genetics research and productivity enhancement. Obstacles include access to patents, commercial licenses and the difficulty that developing countries have in accessing genetic resources and other intellectual property. The International Treaty on Plant Genetic Resources for Food and Agriculture attempted to remedy this problem, but results have been inconsistent. As a result, "orphan crops", such as teff, millets, cowpeas and indigenous plants, which are important in these countries receive little investment. Writing about Norman Borlaug's 2000 publication Ending world hunger: the promise of biotechnology and the threat of antiscience zealotry, the authors argued that Borlaug's warnings were still true in 2010: GM crops are as natural and safe as today's bread wheat, opined Dr. Borlaug, who also reminded agricultural scientists of their moral obligation to stand up to the antiscience crowd and warn policy makers that global food insecurity will not disappear without this new technology and ignoring this reality would make future solutions all the more difficult to achieve.

Yield

US maize yields were flat until the 1930s, when the adoption of conventional hybrid seeds caused them to increase by ~.8 bushels/acre (1937–1955). Thereafter a combination of improved genetics, fertilizer and pesticide availability and mechanization raised the rate of increase to 1.9 bushels per acre per year. In the years since the advent of GM maize, the rate increased slightly to 2.0. Average US maize yields were 174.2 bushels per acre in 2014.

Commercial GM crops have traits that reduce yield loss from insect pressure or weed interference.

2014 Review

A 2014 review, concluded that GM crops' effects on farming were positive. According to The Economist, the meta-analysis considered all published English-language examinations of the agronomic and economic impacts between 1995 and March 2014. The study found that herbicide-tolerant crops have lower production costs, while for insect-resistant crops the reduced pesticide use was offset by higher seed prices, leaving overall production costs about the same.

Yields increased 9% for herbicide tolerance and 25% for insect resistance. Farmers who adopted GM crops made 69% higher profits than those who did not. The review found that GM crops help farmers in developing countries, increasing yields by 14 percentage points. The researchers considered some studies that were not peer-reviewed, and a few that did not report sample sizes. They attempted to correct for publication bias, by considering sources beyond academic journals. The large data set allowed the study to control for potentially confounding variables such as fertiliser use. Separately, they concluded that the funding source did not influence study results.

2010 Review

A 2010 article, supported by CropLife International summarised the results of 49 peer reviewed studies. On average, farmers in developed countries increased yields by 6% and 29% in developing countries. Tillage decreased by 25–58% on herbicide-resistant soybeans. Glyphosate-resistant crops allowed farmers to plant rows closer together as they did not have to control post-emergent weeds with mechanical tillage. Insecticide applications on Bt crops were reduced by 14–76%. 72% of farmers worldwide experienced positive economic results.

2009 Review

In 2009, the Union of Concerned Scientists, a group opposed to genetic engineering and cloning of food animals, summarized peer-reviewed studies on the yield contribution of GM soybeans and maize in the US.The report concluded that other agricultural methods had made a greater contribution to national crop yield increases in recent years than genetic engineering.

Wisconsin study

A study unusually published as correspondence rather than as an article examined maize modified to express four traits (resistance to European corn borer, resistance to corn root worm, glyphosate tolerance and glyfosinate tolerance) singly and in combination in Wisconsin fields from 1990–2010.The variance in yield from year to year was reduced, equivalent to a yield increase of 0.8–4.2 bushels per acre. Bushel per acre yield changes were +6.4 for European corn borer resistance, +5.76 for glufosinate tolerance, "5.98 for glyphosate tolerance and "12.22 for corn rootworm resistance. The study found interactions among the genes in multi-trait hybrid strains, such that the net effect varied from the sum of the individual effects. For example, the combination of European corn borer resistance and glufosinate tolerance increased yields by 3.13, smaller than either of the individual traits

Market dynamics

The seed industry is dominated by a small number of vertically integrated firms. In 2011, 73% of the global market was controlled by 10 companies.

In 2001, the USDA reported that industry consolidation led to economies of scale, but noted that the move by some companies to divest their seed operations questioned the long-term viability of these conglomerates. Two economists have said that the seed companies' market power could raise welfare despite their pricing strategies, because "even though price discrimination is often considered to be an unwanted market distortion, it may increase total welfare by increasing total output and by making goods available to markets where they would not appear otherwise."

Market share gives firms the ability to set or influence price, dictate terms, and act as a barrier to entry. It also gives firms bargaining power over governments in policy making. In March 2010, the US Department of Justice and the US Department of Agriculture held a meeting in Ankeny, Iowa, to look at the competitive dynamics in the seed industry. Christine Varney, who heads the antitrust division in the Justice Department, said that her team was investigating whether biotech-seed patents were being abused.A key issue was how Monsanto licenses its patented glyphosate-tolerance trait that was in 93 percent of US soybeans grown in 2009. About 250 family farmers, consumers and other critics of corporate agriculture held a town meeting prior to the government meeting to protest Monsanto's purchase of independent seed companies, patenting seeds and then raising seed prices.

Intellectual property

Traditionally, farmers in all nations saved their own seed from year to year. However, since the early 1900s hybrid crops have been widely used in the developed world and seeds to grow these crops are purchased each year from seed producers.The offspring of the hybrid corn, while still viable, lose hybrid vigor (the beneficial traits of the parents). This benefit of first-generation hybrid seeds is the primary reason for not planting second-generation seed. However, for non-hybrid GM crops, such as GM soybeans, seed companies use intellectual property law and tangible property common law, each expressed in contracts, to prevent farmers from planting saved seed. For example, Monsanto's typical bailment license (covering transfer of the seeds themselves) forbids saving seeds, and also requires purchasers to sign a separate patent license agreement. Corporations say that they need to prevent seed piracy, to fulfill financial obligations to shareholders, and to finance further development. DuPont spent approximately half its $2 billion research and development (R&D) budget on agriculture in 2011 while Monsanto spends 9–10% of sales on R&D.

Detractors such as Greenpeace say that patent rights give corporations excessive control over agriculture. The Center for Ecoliteracy claimed that "patenting seeds gives companies excessive power over something that is vital for everyone". A 2000 report stated, "If the rights to these tools are strongly and universally enforced - and not extensively licensed or provided pro bono in the developing world - then the potential applications of GM technologies described previously are unlikely to benefit the less developed nations of the world for a long time" (i.e. until after the restrictions expire). Monsanto has patented its seed and it obligates farmers who choose to buy its seeds to sign a license agreement, obligating them store or sell, but not plant, all the crops that they grow.

Besides large agri-businesses, in some instances, GM crops are also provided by science departments or research organisations which have no commercial interests.

LAWSUITS FILED AGAINST FARMERS FOR PATENT INFRINGEMENT

Monsanto has filed patent infringement suits against 145 farmers, but proceeded to trial with only 11.In some of the latter, the defendants claimed unintentional contamination by gene flow, but Monsanto won every case. Monsanto Canada's Director of Public Affairs stated, "It is not, nor has it ever been Monsanto Canada's policy to enforce its patent on Roundup Ready crops when they are present on a farmer's field by accident ... Only when there has been a knowing and deliberate violation of its patent rights will Monsanto act." In 2009 Monsanto announced that after its soybean patent expires in 2014, it will no longer prohibit farmers from planting soybean seeds that they grow.

One example of such litigation is the Monsanto v. Schmeiser case.This case is widely misunderstood.In 1997, Percy Schmeiser, a canola breeder and grower in Bruno, Saskatchewan, discovered that one of his fields had canola that was resistant to Roundup. He had not purchased this seed, which had blown onto his land from neighboring fields. He later harvested the area and saved the crop in the back of a pickup truck. Before the 1998 planting, Monsanto representatives informed Schmeiser that using this crop for seed would infringe the patent, and offered him a license, which Schmeiser refused. According to the Canadian Supreme Court, after this conversation "Schmeiser nevertheless took the harvest he had saved in the pick-up truck to a seed treatment plant and had it treated for use as seed. Once treated, it could be put to no other use. Mr. Schmeiser planted the treated seed in nine fields, covering approximately 1,000 acres in all ... A series of independent tests by different experts confirmed that the canola Mr. Schmeiser planted and grew in 1998 was 95 to 98 percent Roundup resistant." After further negotiations between Schmeiser and Monsanto broke down, Monsanto sued Schmeiser for patent infringement and prevailed in the initial case. Schmeiser appealed and lost, and appealed again to the Canadian Supreme Court, which in 2004 ruled 5 to 4 in

Monsanto's favor, stating that "it is clear on the findings of the trial judge that the appellants saved, planted, harvested and sold the crop from plants containing the gene and plant cell patented by Monsanto".

International trade

GM crops have been the source of international trade disputes and tensions within food-exporting nations over whether introduction of genetically modified crops would endanger exports to other countries.

In Canada in 2010, flax exports to Europe were rejected when traces of an experimental GM flax were found in shipments.This led a member of Parliament to propose Private Member's Bill C-474, which would have required that "an analysis of potential harm to export markets be conducted before the sale of any new genetically engineered seed is permitted". Opponents claimed that "incorporating stringent socio-economic standards into the science-based regulatory system could spell the end of private research funding; because if private biotechnology companies can't see the possibility of a return on their investment, they'll invest their research budget elsewhere".The bill was defeated 176 to 97 in 2011.

REGULATION

Labeling

Green: Mandatory labeling required; Red: Ban on import and cultivation of genetically engineered food. Regulations are pending in Israel and the United States.

In 2014, 64 countries required labeling of all GM foods. These include the European Union,Japan,Australia,New Zealand,Russia,China and India. As of March 2015, Israel was in the process of issuing regulations for labeling of food with ingredients from GMOs. Alaska required labeling of GMO fish and shellfish in 2005, even though no GM fish had been approved by the FDA at the time. A 2014 Vermont law went into effect on July 1, 2016, and some food manufacturers (including General Mills, Mars, Kellogg's, the Campbell Soup Company, PepsiCo, ConAgra, Frito-Lay, and Bimbo Bakeries USA) began distributing products either locally or nationwide with labels such as "Partially produced with Genetic Engineering". Other manufacturers removed about 3,000 non-compliant products from sale in Vermont. The federal government of the United States passed a law at the end of that month pre-empting all state laws, including Vermont's. The law requires labeling regulations to be issued by July 2018, and allows indirect disclosure such as with a phone number, bar code, or web site. It is unclear whether the rules will require labeling of oils and sugars from GM crops, where the final product does not contain any "genetic material" as mentioned in the law.

Prior to the new federal rules taking effect, while it does require pre-market approval, the U.S. Food and Drug Administration has not required GMO labeling as long as there are no differences in health, environmental safety, and consumer expectations based on the packaging. The federal rules come after GMO labeling was debated in many state legislatures and defeated in popular referendums in Oregon (2002 and 2014), Colorado (2014),California Proposition 37 (2012), and Washington Initiative 522 (2012). Connecticut and Maine had passed laws in 2013 and 2014 respectively, which would have required GMO food labels if Northeast states with a population of at least 20 million had passed similar laws (and for Connecticut, representing at least four states).

Other jurisdictions make such labeling voluntary or have had plans to require labeling. Major GM food crop exporters like the United States (until 2018), Argentina, and Canada have adopted voluntary labeling approaches; China and Brazil have major GM (largely non-food) crops and have adopted mandatory labelling.

Arguments

The American Public Health Association, the British Medical Association and the Public Health Association of Australia support mandatory labeling. The European Commission argued that mandatory labeling and traceability are needed to allow for informed choice, avoid potential misleading of consumers and facilitate the withdrawal of products if adverse effects on health or the environment are discovered.A 2007 study on the effect of labeling laws found that once labeling went into effect, few products continued to contain GM ingredients. The study also found that costs were higher in food-exporting than in food-importing countries.

The American Medical Association (AMA) and the American Association for the Advancement of Science have opposed mandatory labeling absent scientific evidence of harm. The AMA said that even voluntary labeling is misleading unless accompanied by focused consumer education. The AAAS stated that mandatory labeling "can only serve to mislead and falsely alarm consumers".

[Labeling] efforts are not driven by evidence that GM foods are actually dangerous. Indeed, the science is quite clear: crop improvement by the modern molecular techniques of biotechnology is safe. Rather, these initiatives are driven by a variety of factors, ranging from the persistent perception that such foods are somehow "unnatural" and potentially dangerous to the desire to gain competitive advantage by legislating attachment of a label meant to alarm. Another misconception used as a rationale for labeling is that GM crops are untested.

Objectivity of regulatory bodies

Groups such as the Union of Concerned Scientists and Center for Food Safety that have expressed concerns about the FDA's lack of a requirement for additional testing for GMO's, lack of required labeling and the presumption that GMO's are "Generally Recognized as Safe" (GRAS), have questioned whether the FDA is too close to companies that seek approval for their products.

Critics in the U.S. protested the appointment of lobbyists to senior positions in the Food and Drug Administration. Michael R. Taylor, a former Monsanto lobbyist, was appointed as a senior adviser to the FDA on food safety in 1991. After leaving the FDA, Taylor became a vice-president of Monsanto. On 7 July 2009, Taylor returned to government as a senior adviser to the FDA Commissioner.

In 2001, when the Starlink corn recall became public, the U.S. Environmental Protection Agency was criticized for being slow to react by Joseph Mendelson III of the Center for Food Safety. He also criticized the EPA and Aventis CropScience for statements at the time of the recall, that indicated they did not anticipate that such a thing would happen. The Canadian Biotechnology Advisory Committee that reviewed Canada's regulations in 2003 was accused by environmental and citizen groups of not representing the full spectrum of public interests and for being too closely aligned to industry groups. Most of the Chinese National Biosafety Committee are involved in biotechnology, a situation that led to criticisms that they do not represent a wide enough range of public concerns.

Litigation and Regulation Disputes

United States

Four federal district courtsuits have been brought against Animal and Plant Health Inspection Service (APHIS), the agency within USDA that regulates genetically modified plants. Two involved field trials (herbicide-tolerant turfgrass in Oregon; pharmaceutical-producing corn and sugar in Hawaii) and two the deregulation of GM alfalfa. and GM sugar beet. APHIS lost all four cases at trial, with the judges ruling they failed to diligently follow the guidelines set out in the National Environmental Policy Act. However, the Supreme Court overturned the nationwide ban on GM alfalfa and an appeal court allowed the partial deregulation of GM sugar beets.After APHIS prepared Environmental Impact Statements for both alfalfa and sugar beets they were approved.

In 2014, Maui County, Hawaii approved an initiative calling for a moratorium on GMO production and research. The initiative specified penalties including fines and jail for knowing violations and did not limit its scope to commercial agriculture. The initiative passed by about 50.2 to 47.9 percent. On December 15, 2015, the New York Times ran an op-ed titled "Are You Eating Frankenfish?", saying that the United States congress will debate whether genetically engineered salmon should be labeled.

European Union

Until the 1990s, Europe's regulation was less strict than in the U.S. In 1998, the use of MON810, a Bt expressing maize conferring resistance to the European corn borer, was approved for commercial cultivation in Europe. However, in the 1990s a series of unrelated food crises created consumer apprehension about food safety in general and eroded public trust in government oversight. A bovine spongiform encephalopathy outbreak was the most publicized.In 1998, a de facto moratorium led to the suspension of approvals of new GMOs in the EU pending the adoption of revised rules. In the mid-1990s, government approval of some GMO crops in the United States precipitated public concern in Europe and led to a dramatic decrease in American exports to Europe. "Prior to 1997, corn exports to Europe represented about 4% of total US corn exports, generating about $300 million in sales ... For example, before 1997, the U.S. sold about 1.75 million tons of corn annually to Spain and Portugal ... But in the 1998–99 crop year, Spain bought less than a tenth of the previous year's amount and Portugal bought none at all."

In May 2003, the US and twelve other countries filed a formal complaint with the World Trade Organization that the EU was violating international trade agreements, by blocking imports of US farm products through its ban on GM food. The countries argued that the EU's regulatory process was far too slow and its standards were unreasonable given the scientific evidence showing that the crops were safe. The case was lobbied by Monsanto and France's Aventis, as well as by US agricultural groups such as the National Corn Growers Association. In response, in June 2003, the European Parliament ratified a U.N.biosafety protocol regulating international trade in GM food, and in July agreed to new regulations requiring labeling and traceability, as well as an opt-out provision for individual countries. The approval of new GMOs resumed in May 2004. While GMOs have been approved since then, approvals remain controversial and various countries have utilized opt-out provisions. In 2006, the World Trade Organization ruled that the pre-2004 restrictions had been violations, although the ruling had little immediate effect since the moratorium had already been lifted. In late 2007, the

US ambassador to France recommended "moving to retaliation" to cause "some pain" against France and the European Union in an attempt to fight the French ban and changes in European policy toward genetically modified crops, according to a US government diplomatic cable obtained by WikiLeaks. 20 out of 28 European Countries (including Switzerland) said No to GMOs until October 2015.

Australia

In May 2014, the Supreme Court of the Australian state of Western Australia dismissed "Marsh v. Baxter". The plaintiff was Steve Marsh, an organic farmer, and the defendant was Michael Baxter, his lifelong neighbour, who grew GM canola. In late 2010, Marsh found seeds from Baxter's crop in his fields. Later, Marsh found escaped GM canola growing amidst his crop. Marsh reported the seed and plants to his local organic certification board, and lost the organic certification of some 70 per cent of his 478 hectare farm. Marsh sued on the grounds that Baxter used a method of harvesting his crop that was substandard and negligent, and on the basis that his land had been widely contaminated. In its summary judgment, the court found that approximately 245 cut canola plants were blown by the wind into Marsh's property, Eagle's Rest. However, Baxter's method (swathing) was "orthodox and well accepted harvest methodology". "In 2011, eight GM canola plants were found to have grown up as self-sown volunteer plants on Eagle Rest", which "were identified and pulled out", and "no more volunteer RR canola plants grew on Eagle Rest in subsequent years". The summary judgment stated that the loss of organic certification "was occasioned by the erroneous application of governing NASAA Standards applicable to NASAA organic operators as regards GMOs (genetically modified organisms) at the time". and that "[t]he absence of a reliable underlying evidentiary platform to support a perpetual injunction against swathing was a significant deficiency".

On June 18, 2014, Marsh announced that he had filed an appeal. One ground was the costs of $803,989 awarded against him. The appeal hearing commenced on 23 March 2015 and was adjourned on 25 March "to deal with an order to ascertain whether Mr Baxter's defence has been financially supported by GM-seed supplier Monsanto and/or the Pastoralists and Graziers Association (PGA)". The Court of Appeal subsequently dismissed the appeal and ordered Marsh to pay Baxter's costs.

Philippines

A petition filed May 17, 2013, by environmental group Greenpeace Southeast Asia and farmer-scientist coalition Masipag (Magsasaka at Siyentipiko sa Pagpapaunlad ng Agrikultura) asked the appellate court to stop the planting of Bt

panel the letters 'GM.'" The rules apply to 19 products including biscuits, breads, cereals and pulses, and a few others. The law faced criticism from consumer rights activists as well as from the packaged-food industry; both sides had major concerns that no logistical framework or regulations had been established to guide the law's implementation and enforcement. On March 21, 2014, the Indian government revalidated 10 GM-based food crops and allowed field trials of GM food crops, including wheat, rice and maize. Food sovereignty", a term coined by members of Via Campesina in 1996, asserts that the people who produce, distribute, and consume food should control the mechanisms and policies of food production and distribution, rather than the corporations and market institutions they believe have come to dominate the global food system. It also encompasses the right of peoples to healthy and culturally appropriate food and their right to define their own food and agriculture systems. The phrase "culturally appropriate" signifies that the food that is available and accessible for the population should fit with the cultural background of the people consuming it.

History

The history of food sovereignty as a movement is relatively young. However, there are a number of key movements and countries that have made significant steps towards making an alternative food system a reality.

Global gatherings

At the Forum for Food Sovereignty in Sélingué, Mali, 27 February 2007, about 500 delegates from more than 80 countries adopted the "Declaration of Nyéléni", which says in part: Food sovereignty is the right of peoples to healthy and culturally appropriate food produced through ecologically sound and sustainable methods, and their right to define their own food and agriculture systems. It puts those who produce, distribute and consume food at the heart of food systems and policies rather than the demands of markets and corporations. It defends the interests and inclusion of the next generation. It offers a strategy to resist and dismantle the current corporate trade and food regime, and directions for food, farming, pastoral and fisheries systems determined by local producers. Food sovereignty prioritises local and national economies and markets and empowers peasant and family farmer-driven agriculture, artisanal fishing, pastoralist-led grazing, and food production, distribution and consumption based on environmental, social and economic sustainability. In April 2008 the International Assessment of Agricultural Science and Technology for Development (IAASTD), an intergovernmental panel under the sponsorship of the United Nations and the World Bank, adopted the following definition: "Food sovereignty is defined as the right of peoples and sovereign states to democratically determine their own agricultural and food policies."

Becoming part of government policy

In September 2008, Ecuador became the first country to enshrine food sovereignty in its constitution. As of late 2008, a law is in the draft stages that is expected to expand upon this constitutional provision by banning genetically modified organisms, protecting many areas of the country from extraction of non-renewable resources, and to discourage monoculture. The law as drafted will also protect biodiversity as collective intellectual property and recognize the Rights of Nature.

Since then another five countries have integrated food sovereignty into their national constitutions or laws. These countries are Venezuela, Mali, Bolivia, Nepal and Senegal; and most recently Egypt (2014 Constitution).

Food sovereignty in Europe

In 2011 more than 400 people from 34 European countries met from the 16th to 21 August in Krems, Austria, to plan the development of a European movement for food sovereignty. The meeting included people from the Atlantic to the Urals and Caucasus, as well as from the Arctic to the Mediterranean. These people included international representatives from diverse social movements and civil society organisations. By coming together they aimed to build on the foundations of the Mali forum in 2007. The objectives were to strengthen local involvement; build a sense of common purpose and understanding; create a joint agenda for action; celebrate the struggle for food sovereignty in Europe; and inspire and motivate people and organisations to work together.

The forum, which was organised on the principles of participation and consensus decision making, used methods to avoid institutionalised prejudices that are inherent in society (such as gender, age, language, occupation). It did this by making a concerted effort to allow for all sections of society to be included in the discussion. The forum allowed producers and activists from projects across Europe to share skills, coordinate actions and discuss perspectives. The forum culminated in the Nyéléni declaration. Since 2011 Europe-wide gatherings and actions have continued, including the Good Food March, where citizens, youth and farmers came together to call for a greener and fairer agricultural policy in Europe, as well as democratic reform of Europe's Common Agricultural Policy.

Food sovereignty versus food security

Food sovereignty was born in response to campaigners' disillusion with food security, the dominant global discourse on food provisioning and policy. The latter emphasises access to adequate nutrition for all, which may be provided by food from one's own country or from global imports. In the name of efficiency and enhanced productivity, it has therefore served to promote what has been termed the "corporate food regime": large-scale, industrialised corporate farming based on specialised production, land concentration and trade liberalisation. Food security's inattention to the political economy of the corporate food regime blinds it to the adverse effects of that regime, notably the widespread dispossession of small producers and global ecological degradation. Haiti can be seen as a case study. Migration from the countryside to cities has reflected a transition from subsistence agriculture to factory labor. Farmers were forced to make this move because of heavy imports of "Miami rice", with which their natively-grown rice could not compete in the local market. By 2008, Haiti was importing 80 percent of its rice, leaving them extremely vulnerable to price and supply fluctuations. When the price of rice did triple in 2008, many Haitians could not afford to buy it. Writing in Food First's Backgrounder, fall 2003, Peter Rosset argues that "food sovereignty goes beyond the concept of food security... [Food security] means that... [everyone] must have the certainty of having enough to eat each day[,] ... but says nothing about where that food comes from or how it is produced." Food sovereignty includes support for smallholders and for collectively owned farms, fisheries, etc., rather than industrializing these sectors in a minimally regulated global economy. In another publication, Food First describes "food sovereignty" as "a platform for rural revitalization at a global level based on equitable distribution of farmland and water, farmer control over seeds, and productive small-scale farms supplying consumers with healthy, locally grown food."

Food justice is a similar concept, but its discourses focuses more on race and class inequities and their relation to food, whereas food sovereignty refers more so to agency over food production systems.

Criticisms of the Green Revolution

The Green Revolution is upheld by some proponents of food security as a success story in increasing crop yields and combating world hunger. However, many in the food sovereignty movement are critical of the green revolution and accuse those who advocate it as following too much of a Western culture technocratic program that is out of touch with the needs of majority of small producers and peasants.

The 'green revolution' refers to developments in plant breeding between the 1960s and 1980s that improved yields from major cereal crops, particularly wheat and rice, and other staple crops. The main focus was on the research, development and transfer of agricultural technology, such as hybrid seeds and fertilisers, through massive private and public investment that went into transforming agriculture in a number of countries, starting in Mexico and India.

While the green revolution may have produced more food, world hunger continues because it did not address the problems of access. Food sovereignty advocates argue that the green revolution failed to alter the highly concentrated distribution of economic power, and if anything, exacerbated it – particularly access to land and purchasing power.

Some critics argue that globally the green revolution caused vast environmental destruction though the increased use of herbicides which caused dramatic loss in biodiversity. There was also a loss of traditional knowledge as farmers relied more heavily of biotechnological inputs. The green revolution favored wealthy, large scale farmers and forced many smaller, poorer farmers into debt.

Some of these views are supported by the World Bank- and UN-sponsored IAASTD report. The focus on technology paid no regard to who controlled that technology and ignored the knowledge of the people who were expected to adopt it. Results included significant biodiversity loss due to the mass adoption of hybrid seeds and soil erosion. The adoption of genetically modified (GMOs) cropping by the government of Western Australia in 2010 and the subsequent failure of crop segregation has led to the contamination of at least one organic farm by Monsanto's GMcanola. The organic certification of the certified organic farm of Steve and Sue Marsh was withdrawn in 2010 due to GMcontamination. A court case in the Supreme Court of Western Australia for nuisance and negligence failed to achieve any relief or protection for the organic farm. However, the organic certification of the Marsh farm was reinstated in 2013.

Academic Perspectives

Food Regime Theory

It is in its capacity as a social movement that food regime analysts are interested in food sovereignty. With its Marxist influences, food regime theorists are interested in how moments of crisis within a particular food regime are expressive of the dialectical tension that animates movement between such configurations (i.e., periods of transition). According to leading theorist Philip McMichael, food regimes are always characterised by contradictory forces. Consolidation of a regime does not so much resolve as it does contain, or else strategically accommodate, these tensions; meanwhile, their intensification, often via the mobilisations of social movements, often signals a period of transition.

According to McMichael, a "world agriculture" under the WTO Agreement on Agriculture ("food from nowhere") represents one pole of the "central contradiction" of the present regime. He is interested in the food sovereignty movement's potential to escalate the tension between this and its opposing pole, the agroecology-based localism ("food from somewhere") advocated by various grassroots food movements. Offering slightly different conclusions, recent work by Harriet Friedmann suggests that "food from somewhere" is already being co-opted under an emergent "corporate-environmental" regime (cf. Campbell 2009).

CRITICISMS

Wrong Baseline Assumptions

Some scholars argue that the Food Sovereignty movement follows wrong baseline assumptions (small-scale farming is not necessarily a freely chosen life-style and farmers in least developed and highly developed countries do not face the same challenges). The Food Sovereignty movement may be right about the mistakes of neoliberal economic ideology, but it is silent about the fact that many famines actually occurred under socialist and communist regimes that pursued the goal of food self-sufficiency (cf. Aerni 2011).

Political-Jurisdictional Model

There is a lack of consensus within the food sovereignty movement regarding the political or jurisdictional community at which its calls for democratisation and renewed "agrarian citizenship" (cf. Wittman 2009) are directed. In public statements, the food sovereignty movement urges for strong sovereign powers for both national governments and local communities (in the vein of the indigenous rights movement, Community-Based Natural Resource Management (CBNRM) and the like) (elsewhere it has also appealed to global civil society to act as a check against abuses by national and supranational governing bodies).

Those who take a radically critical view on state sovereignty would argue against the possibility that national sovereignty can be reconciled with that of local communities (see also the debate about multiculturalism and indigenous autonomy in Mexico). On the other hand, Raj Patel is more favourable towards this prospect: for Patel, an adapted version of Seyla Benhabib's Kantian-inspired cosmopolitan federalism – involving multiple, layered geographies of democratic attachment and jurisdiction – could offer a promising vehicle for the realisation of food sovereignty on a large scale. Patel's important proviso here is that a stronger version of Benhabib's accompanying principle of moral universalism is also pursued. By Patel's assessment, the food sovereignty movement is showing promising signs of moving towards the radical egalitarianism and democratic praxis that such a model entails.

Crisis of the peasantry?

In its strong reassertion of rural and peasant identities and forms of social reproduction, the food sovereignty movement has been read as a sharp challenge to modernist narratives of inexorable urbanisation, industrialisation of agriculture and de-peasantisation. However, as part of ongoing debates over the contemporary relevance of the "agrarian question" in classical Marxism, Henry Bernstein is critical of these largely celebratory accounts. Specifically, levels Bernstein, such analyses tend to present the agrarian population as a unified, singular and world-historical social category. Therefore, they fail to account, in any more than a gestural manner, for:

- this population's vast internal social differentiation (North/South, gender and class positionalities);
- the conservative, cultural survivalist tendencies of a movement that has emerged as part of a backlash against the perceived homogenising forces of globalisation (Boyer discusses whether food sovereignty is a counter or anti-development narrative

7

INSECT RESISTANCE HORTICULTURAL PLANT

Like disease, insects are important causal factors of biotic stress in crop plants. Insects attack all the crop plants and lead to considerable losses in yield as well as quality. Insect attack leads to various types of damages. In crop plants such as 1) Reduction in plant growth or stunting, 2) Damage of vegetative and reproductive parts, 3) Premature defoliation, and 4) Wilting of plants. Insect cause 14% estimated yield loss of all important crops on global bases. Insect cause yield loss directly either by sucking cell sap or by eating away various plant parts. Insects also cause yield loss through transmission of various diseases.

There are two important methods of insect control, viz 1) biological method and 2) chemical method. In biological method, insects are controlled in three ways, viz, 1) by the use of predators and parasites of insect pests. 2) by using botanical pesticides such as neem, Datura, Ipomea in the form of leaf extracts, and 3) use of resistant varieties. Biological method is cheap and does not have any adverse effect on the ecosystem, through lesser effective than chemical method of insect control. The chemical method includes use of various chemical, insecticides. Use of insecticides has several disadvantages. It increases cost of cultivation, reduces population of predators and parasites of insect pests, leads to environmental pollution and development of pesticides resistant biotypes of insects. Thus genetic resistance is the cheapest and the best method of insect control in crop plants. Genetic resistance refers to the ability of some genotypes to give higher yields of good quality than susceptible varieties at the same initial level of insect attack under similar environmental conditions. Thus resistance is defined in relation to susceptible varieties. This chapter deals with breeding for resistance to insect pests and parasitic weeds.

MECHANISMS OF INSECT RESISTANCE

There are four mechanisms of insect resistance, viz 1) no preference , 2) antibiotics, 3) tolerance, and 4) avoidance or escape. The first three mechanisms were given by Painter (1951) and the fourth one was added subsequently. A resistant variety may have one, two or more of these mechanisms.

The Bt Chapter

Bacillus thuringiensis, commonly known as Bt, is a bacterium that occurs naturally in the soil. For years, bacteriologists have known that some strains of Bt produce proteins that kill certain insects with alkaline digestive tracts. When these insects ingest the protein produced by Bt, the function of their digestive systems is disrupted, producing slow growth and, ultimately, death.

Bt is very selective - different strains of the bacterium kill different insects and only those insects. Strains of Bt are effective against European corn borers and cotton bollworms (Lepidoptera), Colorado potato beetles (Coleoptera), and certain flies and mosquitos (Diptera). Bt is not harmful to humans, other mammals, birds, fish, or beneficial insects.

Bt was first identified in 1911 when it was discovered that it killed the larvae of flour moths. Bt was registered as a biopesticide in the U.S. in 1961. Today it is used in insecticide sprays sold to home gardeners and others worldwide.

In 1983, the World Health Organization used Bt in West Africa to control disease-carrying blackflies. In the U.S., various strains of Bt are used to control spruce budworms and gypsy moths in forests, cabbage worms in broccoli and cauliflower, loopers or budworms in cotton and tobacco, and leaf rollers in fruits.

However, less than one percent of all pesticides used in the U.S. each year contain Bt (Monsanto). As an ingredient of commercial sprays, Bt is relatively expensive and has some drawbacks. Although some pesticides kill on contact, Bt must be eaten by insects to be effective. Sunlight breaks down Bt, and rain washes it from the plants. Therefore, Bt must be applied exactly where and when the target insects are feeding and they must consume it quickly before it disappears.

Bt and Biotechnology

Today, plants can be genetically engineered to produce their own Bt. Genetic (recombinant DNA) engineering is the modification of DNA molecules to produce changes in plants, animals, or other organisms. DNA (deoxyribonucleic acid) is a double-stranded molecule that is present in every cell of an organism and contains the hereditary information that passes from parents to offspring. This hereditary information is contained in individual units or sections of DNA called genes. The genes that are passed from parent to offspring determine the traits that the offspring will have.

In the last twenty years, scientists made a surprising discovery - DNA is interchangeable among animals, plants, bacteria ... any organism. In addition to using traditional breeding methods of improving plants and animals through years of crossbreeding and selection, scientists can now isolate the gene or genes for the traits they want in one animal or plant and move them into another. Of course, when a trait is controlled by several genes, the transfer process is more difficult. The plants or animals modified in this way are called transgenic.

Out of Bt. . . Into the Plants

DNA technology makes it possible to locate the gene that produces Bt proteins lethal to insects and transfer the gene into crop plants. The process is depicted in.

First, scientists identify a strain of Bt that kills the targeted insect. Then they isolate the gene that produces the lethal protein. That gene is removed from the Bt bacterium and a gene conferring resistance to a chemical (usually antibiotic or herbicide) is attached that will prove useful in a later step.

The Bt gene with the resistance gene attached is inserted into plant cells. At this point, scientists must determine which plant cells have successfully received the Bt gene and are now transformed. Any plant cell that has the Bt gene must also have the resistance gene that was attached to it. Researchers grow the plant cells in the presence of the antibiotic or herbicide and select the plant cells that are unaffected by it. These genetically transformed plant cells are then grown into whole plants by a process called tissue culture. The modified plants produce the same lethal Bt protein produced by Bt bacteria because the plants now have the same gene.

Research to transfer insect resistance genes from Bt to crop plants is well under way. Corn, cotton, and potatoes are three of the many commercial crops targeted for Bt insect resistance.

Bt in Corn Confers Resistance to European Corn Borers

Modern corn hybrids have some resistance to European corn borers (ECB), but they still sustain damage from moderate and high levels of infestation. Genetically engineered insect-resistant corn is providing a new defense against an old enemy.

The Enemy

The European corn borer is a major pest of corn in North America and Europe. Researchers estimate that yield loss due to ECB averages five percent, but any one year or location can have much higher damage (Ciba Seeds). The number of larvae in the previous season, the type of winter weather, natural parasites and disease, and weather conditions during the growing season can affect ECB levels. For example, during the 1990 and 1991 growing seasons, ECB levels were very high across much of the U.S. Corn Belt and losses were heavy. But the cool, rainy weather and diseases during critical parts of the 1992 growing season reduced ECB populations and their damage in many areas.

ECB larvae attack corn two times during the growing season across most of the U.S. Corn Belt. First-brood larvae feed on plant leaves before flowering, injure leaf tissue, and eventually bore into the stalk. Second-brood larvae begin feeding in the leaf sheath and collar area of corn plants after flowering. Eventually, they bore into the stalk and ear shank where they can cause broken stalks and dropped ears.

The greatest source of ECB damage may not be visible. ECB disrupts the physiological processes of corn and can reduce yields even when stalk lodging (breaking) problems are not evident. Physiological damage can cause low test weight of harvested corn, small ears, or early death of the plant.

Chemical sprays must be applied before the borer tunnels into the stalk to be effective. Multiple applications may be required, and, even then, complete control is not possible.

The Bt Solution

Scientists know that a family of insecticidal proteins from Bt kills ECB. The bacteria produce the protein as an insoluble crystal. When a susceptible caterpillar, like the corn borer, eats the crystal, part of it binds to, penetrates, and collapses the cells lining its gut, causing death.

Although Bt genes have been introduced into tobacco, tomatoes, cotton, and other broadleaf plants, gene transfer technology for corn is a recent achievement. The development of corn plants expressing Bt proteins requires substantial changes in the Bt genes, including the creation of synthetic versions of the genes, rather than the microbial Bt gene itself. For example, one company has developed three versions of a synthetic Bt gene that "switches" on in pollen, in green tissue, or in other parts of the corn plant (Ciba Seeds). Along with a herbicide resistance gene, the three Bt versions were "shot" by a biolistic gene gun into immature corn embryos taken from developing seeds about 15 days after pollination. The gene gun bombarded the embryos with tiny gold particles coated with the genes.

Some plants received genes with a non-specific promoter (switch) that resulted in production of the Bt protein in many parts of the corn plant. Other plants received a mixture of genes with switches that resulted in Bt protein production only in the pollen and green tissue. The herbicide resistance gene allowed scientists to determine which plant cells had received a gold particle carrying DNA and successfully incorporated the gene. Subsequent screening for Bt production showed which cells received both herbicide resistance and the Bt gene(s).

When the cells were grown into plants and field tested for resistance, it was found that the Bt plants fared better than any of the control plants. In general, plants with both the pollen and green tissue Bt genes appeared more resistant to second-brood corn borer damage than plants with Bt genes that expressed themselves in many parts of the plant. Another advantage of the Bt genes with specific switches (pollen and green tissue) was that they produced insecticidal protein in those parts of the plant attacked by both first-and secondbrood ECBs, while minimizing production in seed and other parts of the plant where protection is not critical.

Using Bt technology to control corn borers is an international effort. One U.S. company is using a biolistic gene gun to bombard Indonesian corn with Bt genes that produce proteins lethal to the Asian corn borer (Wilson). This pest is responsible for significant yield losses, up to 40 percent, in much of the corn-growing areas of Indonesia.

The warm Indonesian climate and the life cycle of the Asian corn borer allow 9 to 12 overlapping generations annually, making it a threat to corn virtually year-round. Introducing Bt insect resistance genes into commercial Indonesian corn seems to be the most feasible and effective way available to control the pest.

Comprehensive testing must be done, approval by relevant authorities obtained, and patent questions answered before any company can commercialize corn hybrids containing the Bt gene.

Bt Resistance for the Colorado Potato Beetle

The potato is the most popular vegetable among North American consumers. It is estimated that potatoes, in some form, are consumed during one of every three meals (Monsanto). Despite their popularity, potatoes are not that easy to grow. Each year, potato growers lose a significant amount of their crops to an insect called the Colorado potato beetle. The beetle feeds on the growing plant leaves and stems during the growing season, stunting the plant and cutting yields. Repeated applications of chemical sprays are usually needed to control the pest.

The Economic Picture

According to the USDA, potato producers in the U.S. spend $20 to $40 million each year for control of the Colorado potato beetle (Monsanto). In an average year, growers spray their fields with insecticides as many as five times at a cost of up to $150 per acre.

If insect-resistant plants were available, researchers estimate that the use of chemical insecticides in potatoes could be reduced as much as 30 to 40 percent. Translated into dollars, growers could save $6 to $16 million annually, based on USDAestimates of current insecticide use.

SIMILAR SCIENCE, DIFFERENT TARGET

The science being used to develop potatoes resistant to the Colorado potato beetle is similar to the European corn borer technology. One of the advantages of Bt is the diversity of strains that can be used to target different insect pests.

Like the corn researchers, potato scientists are using a specific type of the naturally occurring soil bacterium Bacillus thuringiensis. A single gene from the Bt is spliced into a chromosome of a potato cell. When the cell carrying the Bt gene is grown into a plant, the plant produces a protein that is toxic to the Colorado potato beetle. When the beetle feeds on the genetically improved plant, the toxic protein interferes with its digestive system and it dies.

The protein is highly selective, affecting only the Colorado potato beetle. It does not harm humans, animals, or beneficial insects that help control other crop pests.

The Outlook

One company conducting insect-resistant potato research expects farmers to be planting their beetle-resistant potatoes by the mid-1990s (Monsanto news release). This company began its development of insect-resistant potatoes with the Russet Burbank variety, since it comprises approximately 40 percent of U.S. potato acreage. However, company researchers are also working to improve other popular potato varieties such as Atlantic, Superior, Norchip, and Shepody.

Like all genetically improved food crops, the exact timetable for commercialization of insectresistant potatoes depends not only on the scientific tests that must be conducted, but also on approvals by the appropriate regulatory agencies.

The potato plant on the left did not produce the Bt protein to protect it form the Colorado potato beetle. The plant on the right had the Bt gene.

Bt Resistance for Cotton

Cotton is another cash crop benefiting from Bt insect-resistance research. With a market value of $4.5 billion, cotton is the nation's fifth largest crop (Monsanto). Yet, annually cotton growers expect to lose up to 15 percent of each field due to insect damage.

Caterpillar insects, like the tobacco budworm, cotton bollworm, and pink bollworm, feed on cotton plants. Experts say that these and similar pests are responsible for millions of dollars in damage to the American cotton crop each year.

Caterpillar Economics

Caterpillar insects are responsible for 60 to 70 percent of all insect damage to cotton plants (Monsanto). Efforts to control them chemically account for about 60 to 70 percent of a cotton grower's pesticide costs. In the Mississippi Delta growing region, a cotton grower's pesticide costs can average $40 to $45 per acre each season. In Arizona, pest control costs are usually higher, perhaps double that of the Mississippi Delta.

Chemical sprays containing Bt are seldom practical in cotton, since Bt breaks down in sunlight, washes away in rain, and is nearly impossible to apply to the plant parts where insects feed. However, other chemical insecticides are used that may need to be applied as many as 10 times in one season. By planting insectresistant cotton, researchers hope farmers can reduce the number of insecticide applications they must pay for.

Same Song, Third Verse

Creating insect resistance in cotton by using Bt technology is a procedure similar to that used in corn or potatoes. Researchers find one or more strains of Bt that produce a protein fatal to some of the most damaging cotton insect pests. The gene that regulates the protein's production is inserted into the genetic structure of cells from a cotton plant. The cells that successfully receive the Bt genes are grown into whole plants through tissue culture and enter a long period of laboratory and field testing. When certain caterpillar pests feed on the new cotton plants, the protein damages their digestive systems, and they die.

Commercialization Timetable

Current estimates are that it will be several years before insectresistant cotton plants are commercially marketed. They must meet the regulatory requirements administered by the USDA and the Environmental Protection Agency (EPA).

THE INSECT RESISTANCE PROBLEM

Transgenic Bt plants like corn, potatoes, and cotton have a possible problem - what if insects build up a resistance to the lethal proteins?

The potential of pests to develop resistance against the defense mechanisms of crops is well-known and is not unique to genetically engineered plants. Insects may develop resistance to a crop defense no matter how it was developed. The crop defense might be a chemical or biological agent, a gene already in the crop species and transferred to commercial plants by conventional plant breeding methods, or a gene introduced by recombinant DNA technology. Because more than 500 insects and mites already have acquired resistance to a number of insecticides, there is concern that similar resistance to Bt toxins could develop (McGaughey and Whalon).

Several major pests, including the tobacco budworm, Colorado potato beetle, Indianmeal moth, and diamondback moth, have demonstrated the ability to adapt to Bt in the laboratory. It has been reported that the diamondback moth evolved high levels of resistance in the field as a result of repeated use of Bt (McGaughey and Whalon). As Bt use increases on more acres, some scientists predict that insect resistance to Bt will be a major problem. Considerable controversy exists about how Bt should be managed to prolong its usefulness.

Bt Durability in Non-Transgenic Crops

The crystal proteins of Bt that are lethal to insects have been widely used as microbial insecticides in traditional corn and vegetable crops for years. In recent years, field populations of insects tolerant to these Bt proteins have been reported. To date, these cases are limited and have been associated with frequent and prolonged use of Bt on geographically isolated insect populations (Ciba Seeds). Scientists have been able to induce higher levels of resistance in these field populations by taking them into the laboratory and exposing them to specific Bt insecticidal crystal proteins.

A laboratory population of tobacco budworm that is resistant to more than one class of crystal proteins has been reported. However, the resistance of this population is not high and its ability to survive is significantly reduced when the proteins are absent.

In the case of ECB-resistant corn, it is thought that the development of insect resistance may be delayed by the availability of non-Bt corn fields (refuges) that are still susceptible to the borers. Borers that could not survive ECB-resistant fields may do well in the refuges. There they could breed with resistant borers, diluting the resistance trait in the gene pool.

The large boll of cotton on the left was produced by a plant designed with a Bt gene to fent of insects. The smaller boll on the right was grown in the same field on an unprotected plant.

Bt Durability in Transgenic Crops

About 13 U.S. and European companies that fund academic Bt research have formed a Bt Management Working Group to support research on the interaction of transgenic plants with target pests.

The plant industry is employing several approaches to deal with insect resistance to transgenic insecticidal plants, including the following:

Very high levels of insecticidal crystal proteins (ICPs) are being produced in new genetically modified plants. The only insects that can survive feeding on such plants will be those that already possess a high-level resistance gene as part of their genetic code. Such genes are expected to be extremely rare in the insect population. Insects with more common genes for partial, or lowlevel, resistance will not survive the very high levels of ICPs.

Additional insecticidal genes are being sought. The idea is to develop transgenic plants that express several insecticidal genes targeting different sites within the insect. In order to survive plants with such multiple defenses, an insect would probably have to possess multiple resistance genes. Some scientists believe it is highly unlikely that insects can acquire such genes. In any case, these scientists anticipate that plants expressing additional insecticidal genes will be available before insect resistance to a single, highly expressed ICP from Bt develops.

Regulating Bt Plants

Insect-resistant crops developed through Bt technology are subject to the same regulations as other genetically engineered plants. It should be noted that U.S. regulatory policy for genetically engineered plants has evolved over time and additional changes in the requirements discussed below may occur. Three government agencies share responsibility for administering plant biotechnology regulations in the United States.

U.S. Department of Agriculture (USDA)

The USDA administers the Federal Plant Pest Act (FPPA), which regulates interstate movement, importation, and field testing of genetically engineered plants. A special permit from the USDA and approval by the individual state departments of agriculture are required to move any genetically engineered organism into the U.S. or between states. Applicants for a permit must provide details about the nature of the organism, its origin, and its intended use. If granted, the permit certifies that the applicant's facility and procedures meet certain operating standards, such as appropriate levels of containment to prevent accidental escape of the organism. Guidelines set by the National Institutes of Health for laboratory research involving recombinant DNA molecules are included in these standards and must be followed by the applicant.

The USDA oversees field testing of genetically engineered crops. The applicant must provide complete information about the plant, including all new genes and new gene products, their origin, the purpose of the test, the experimental design (how the test will be conducted), and precautions to be taken to prevent the escape of pollen, plants, or plant parts from the field test site. The approval process can take a maximum of four months.

Before a genetically engineered crop can be sold commercially, a petition must be filed for USDA exemption. This petition requires more information than a field test permit, including environmental product safety information.

Recently, the USDA streamlined the permit process for movement and field testing of certain genetically engineered crops such as corn, soybeans, cotton, potatoes, tomatoes, and tobacco. The streamlining was based on an extensive history of safe use in field trials in the United States. In the streamlined process, an applicant can field test specific crops if the crop and genetic material meet certain USDA performance standards. For a field test that does not meet these performance standards, the current 120-day field test permit process would be followed.

Environmental Protection Agency (EPA)

The EPA recently proposed new policies for review of applications for field trials of genetically engineered plants. It proposed that substances introduced into crops by genetic engineering to provide disease or insect resistance (Bt crops are in this category) fall under the Federal Insecticide, Fungicide, and Rodenticide Act and the Federal Food, Drug, and Cosmetic Act. To date, the USDA has consulted with EPA when reviewing applications to test such crops on a small scale. The EPA will most likely formally review a product when a company decides to move into large-scale testing. As with traditional pesticides, an Experimental Use Permit (EUP) will be required for tests that exceed 10 acres total in the U.S.

After receiving the EUP, a company follows a process that involves extensive testing of the crop to ensure food and environmental safety before the crop is commercialized. In addition, it is likely that some applicants also will consult the Food and Drug Administration (FDA) before a pest-resistant food crop is commercialized, even though a recent FDA policy proposal emphasized that such products fall under the EPA's jurisdiction and are not subject to FDA pre-market approval.

Food and Drug Administration (FDA)

The FDA also has very broad authority under the Federal Food, Drug, and Cosmetic Act to regulate the introduction of new food crops, whether they were conventionally produced or created by genetic engineering. FDA policy states that any new food developed through genetic modification may require testing and/or labeling if the genetic material introduced into the host plant comes from outside the traditional food supply, is a known or suspected cause of allergic reactions, or significantly alters the nutritional composition of the food (Federal Register). Every company or individual that produces food or food products through recombinant DNA technology is legally required to assure its safety and quality before it enters the food supply - the same assurance required of traditional food products.

The Bt Outlook

Crops that have been genetically engineered for Bt resistance could dramatically lower production costs and provide farmers with new insect control options within the next few years. The success of their commercialization depends on several factors, including the regulatory climate, patent issues, and the ability of scientists to deal with targeted insects that develop resistance to the lethal proteins. Several companies currently working on Bt insect resistance in crops foresee marketing their products in the mid1990s.

8

DISEASE RESISTANCE HORTICULTURAL PLANT

The method of breeding for disease resistance is essentially the same as those for other agronomic characters. The following breeding methods have commonly used, 1) Selection, 2) Introduction, 3) Mutation, 4) Hybridization, 5) Somaclonal Variation, and 6) Genetic engineering.

INTRODUCTION

This is easy and rapid method of developing disease resistant variety. The resistant variety may be introduced and after testing, if found suitable, can be released in the disease prone area. In 1860, the grape crop in France was completely destroyed by the attack of Phylloxera Vertifolia. Introduction of resistant root stocks to this pest from USA saved the grape crop from extinction in France.

Selection

When the source of resistance is a cultivated variety, mass selection and pure lines selection in self pollinated crops, mass and recurrent selection in cross pollinated species, and clonal selection in the vegetatively propagated crops will be ideal for isolating disease resistant plants. The resistant plants may be multiplied, screened for disease resistance and released a variety.

Hybridization

Hybridization is used when resistant genes are available either in the germplasm or in wild species of crop plants. After hybridization, the hybrid material is handled either by pedigree method or by backcross method. The pedigree method is used when the resistance is governed by polygene and the resistant variety is an adapted one which also contributes some desirable agronomic traits. The backcross method is used when resistance to governed by oligogenes.

Induced mutations are also use for disease resistance. Many disease resistant varieties have been developed in various crops through induced mutations.

Mutation

We have already considered briefly the usefulness of spontaneous as well as induced.

Somaclonal Variation

Disease resistant soma clonal variants can be obtained in the following two ways, firstly , plants regenerated from cultured cells or their progeny are subjected to disease test and resistant plants are isolated. Secondly, cultured cells are selected for resistance to the toxin or culture filterate produced by the pathogen and plants are regenerated from the selected cell. In most cases, these plants are also resistant to the disease in question. Cell selection strategy is most likely to be successful in cases where the toxin is involved in disease development.

Genetic Engineering

Genes expected to confer disease resistance are isolated, cloned and transferred into the crop in question. In case of viral pathogens, several transgenes have been evaluated, viz, virus coat protein gene, DNA copy of viral satellite RNA, defective viral genome, antisense constructs of critical viral genes, and ribozymes. Viral coat protein gene approach seems to be the most successful. A virus transgenic variety of squash is in commercial cultivation in U.S.A.

Plant disease resistance

Plant disease resistance protects plants from pathogens in two ways: by pre-formed structures and chemicals, and by infection-induced responses of the immune system. Relative to a susceptible plant, disease resistance is the reduction of pathogen growth on or in the plant, while the term disease tolerance describes plants that exhibit little disease damage despite substantial pathogen levels. Disease outcome is determined by the three-way interaction of the pathogen, the plant and the environmental conditions (an interaction known as the disease triangle).

Defense-activating compounds can move cell-to-cell and systemically through the plant vascular system. However, plants do not have circulating immune cells, so most cell types exhibit a broad suite of antimicrobial defenses. Although obvious qualitative differences in disease resistance can be observed when multiple specimens are compared (allowing classification as "resistant" or "susceptible" after infection by the same pathogen strain at similar inoculum levels in similar environments), a gradation of quantitative differences in disease resistance is more typically observed between plant strains or genotypes. Plants consistently resist certain pathogens but succumb to others; resistance is usually specific to certain pathogen species or pathogen strains.

Background

Plant disease resistance is crucial to the reliable production of food, and it provides significant reductions in agricultural use of land, water, fuel and other inputs. Plants in both natural and cultivated populations carry inherent disease resistance, but this has not always protected them.

The late blight Irish potato famine of the 1840s was caused by the oomycete Phytophthora infestans. The world's first mass-cultivated banana cultivar Gros Michel was lost in the 1920s to Panama disease caused by the fungus Fusarium oxysporum. The current wheat stem, leaf, and yellow stripe rust epidemics spreading from East Africa into the Indian subcontinent are caused by rust fungi Puccinia graminis and P. striiformis. Other epidemics include Chestnut blight, as well as recurrent severe plant diseases such as Rice blast, Soybean cyst nematode, Citrus canker.

Plant pathogens can spread rapidly over great distances, vectored by water, wind, insects, and humans. Across large regions and many crop species, it is estimated that diseases typically reduce plant yields by 10% every year in more developed nations or agricultural systems, but yield loss to diseases often exceeds 20% in less developed settings, an estimated 15% of global crop production.

However, disease control is reasonably successful for most crops. Disease control is achieved by use of plants that have been bred for good resistance to many diseases, and by plant cultivation approaches such as crop rotation, pathogen-free seed, appropriate planting date and plant density, control of field moisture and pesticide use.

Common Disease Resistance Mechanisms

Pre-formed Structures and Compounds

- Plant cuticle/surface
- Plant cell walls

- Antimicrobial chemicals (for example: glucosides, saponins)
- Antimicrobial proteins
- Enzyme inhibitors
- Detoxifying enzymes that break down pathogen-derived toxins
- Receptors that perceive pathogen presence and activate inducible plant defences

Inducible post-infection Plant Defenses

- Cell wall reinforcement (callose, lignin, suberin, cell wall proteins)
- Antimicrobial chemicals, including reactive oxygen species such as hydrogen peroxide or peroxynitrite, or more complex phytoalexins such as genistein or camalexin
- Antimicrobial proteins such as defensins, thionins, or PR-1
- Antimicrobial enzymes such as chitinases, beta-glucanases, or peroxidases
- Hypersensitive response - a rapid host cell death response associated with defence mediated by "Resistance genes."(Bryant, Tracy 2008).

Immune system

The plant immune system carries two interconnected tiers of receptors, one most frequently sensing molecules outside the cell and the other most frequently sensing molecules inside the cell. Both systems sense the intruder, respond to the intrusion and signal to the rest of the plant and sometimes to neighboring plants that the intruder is present. The two systems detect different types of pathogen molecules and classes of plant receptor proteins.

The first tier is primarily governed by pattern recognition receptors that are activated by recognition of evolutionarily conserved pathogen or microbial–associated molecular patterns (PAMPs or MAMPs). Activation of PRRs leads to intracellular signaling, transcriptional reprogramming, and biosynthesis of a complex output response that limits colonization. The system is known as PAMP-Triggered Immunity or as Pattern-Triggered Immunity (PTI).

The second tier, primarily governed by R gene products, is often termed effector-triggered immunity (ETI). ETI is typically activated by the presence of specific pathogen "effectors" and then triggers strong antimicrobial responses.

In addition to PTI and ETI, plant defenses can be activated by the sensing of damage-associated compounds (DAMP), such as portions of the plant cell wall released during pathogenic infection. Responses activated by PTI and ETI receptors include ion channel gating, oxidative burst, cellular redox changes, or protein kinase cascades that directly activate cellular changes (such as cell wall reinforcement or antimicrobial production), or activate changes in gene expression that then elevate other defensive responses

Plant immune systems show some mechanistic similarities with the immune systems of insects and mammals, but also exhibit many plant-specific characteristics. The two above-described tiers are central to plant immunity but do not fully describe plant immune systems. In addition, many specific examples of apparent PTI or ETI violate common PTI/ETI definitions, suggesting a need for broadened definitions and/or paradigms.

PAMP-triggered Immunity

PAMPs, conserved molecules that inhabit multiple pathogen genera, are referred to as MAMPs by many researchers. The defenses induced by MAMP perception are sufficient to repel most pathogens. However, pathogen effector proteins are adapted to suppress basal defenses such as PTI. Many receptors for MAMPs (and DAMPs) have been discovered. MAMPs and DAMPs are often detected by transmembrane receptor-kinases that carry LRR or LysM extracellular domains.

Effector Triggered Immunity

Effector Triggered Immunity (ETI) is activated by the presence of pathogen effectors. The ETI response is reliant on R genes, and is activated by specific pathogen strains. Plant ETI often causes an apoptotic hypersensitive response.

R genes and R proteins

Plants have evolved R genes (resistance genes) whose products mediate resistance to specific virus, bacteria, oomycete, fungus, nematode or insect strains. R gene products are proteins that allow recognition of specific pathogen effectors, either through direct binding or by recognition of the effector's alteration of a host protein. Many R genes encode NB-LRR proteins (proteins with nucleotide-binding and leucine-rich repeat domains, also known as NLR proteins or STAND proteins, among other names). Most plant immune systems carry a repertoire of 100-600 different R gene homologs. Individual R genes have been demonstrated to mediate resistance to specific virus, bacteria, oomycete, fungus, nematode or insect strains. R gene products control a broad set of disease resistance responses whose induction is often sufficient to stop further pathogen growth/spread.

Studied R genes usually confer specificity for particular strains of a pathogen species (those that express the recognized effector). As first noted by Harold Flor in his mid-20th century formulation of the gene-for-gene relationship, a plant R gene has specificity for a pathogen avirulence gene (Avr gene). Avirulence genes are now known to encode effectors. The pathogen Avr gene must have matched specificity with the R gene for that R gene to confer resistance, suggesting a receptor/ligand interaction for Avr and R genes. Alternatively, an effector can modify its host cellular target (or a molecular decoy of that target), and the R gene product (NLR protein) activates defenses when it detects the modified form of the target or decoy.

Effector biology

Effectors are central to the pathogenic or symbiotic potential of microbes and microscopic plant-colonizing animals such as nematodes. Effectors typically are proteins that are delivered mostly outside the microbe and into the host cell. Effectors manipulate cell physiology and development. As such, effectors offer examples of co-evolution (example: a fungal protein that functions outside of the fungus but inside of plant cells has evolved to take on plant-specific functions). Pathogen host range is determined, among other things, by the presence of appropriate effectors that allow colonization of a particular host. Pathogen-derived effectors are a powerful tool to identify host functions that are important in disease. Apparently most effectors function to manipulate host physiology to allow disease to occur. Well-studied bacterial plant pathogens typically express a few dozen effectors, often delivered into the host by a Type III secretion apparatus. Fungal, oomycete and nematode plant pathogens apparently express a few hundred effectors.

So-called "core" effectors are defined operationally by their wide distribution across the population of a particular pathogen and their substantial contribution to pathogen virulence. Genomics can be used to identify core effectors, which can then be used to discover new R gene alleles, which can be used in plant breeding for disease resistance.

Against viruses, plants often induce pathogen-specific gene silencing mechanisms mediated by RNA interference. This is a form of adaptive immunity.

Plant immune systems also can respond to an initial infection in one part of the plant by physiologically elevating the capacity for a successful defense response in other parts. Such responses include systemic acquired resistance, largely mediated by salicylic acid-dependent pathways, and induced systemic resistance, largely mediated by jasmonic acid-dependent pathways.

Species-level resistance

In a small number of cases, plant genes are effective against an entire pathogen species, even though that species that is pathogenic on other genotypes of that host species. Examples include barley MLO against powdery mildew, wheat Lr34 against leaf rust and wheat Yr36 against stripe rust. An array of mechanisms for this type of resistance may exist depending on the particular gene and plant-pathogen combination. Other reasons for effective plant immunity can include a lack of coadaptation (the pathogen and/or plant lack multiple mechanisms needed for colonization and growth within that host species), or a particularly effective suite of pre-formed defenses.

SIGNALING MECHANISMS

Perception of pathogen presence

Plant defense signaling is activated by the pathogen-detecting receptors that are described in an above section. The activated receptors frequently elicit reactive oxygen and nitric oxide production, calcium, potassium and proton ion fluxes, altered levels of salicylic acid and other hormones and activation of MAP kinases and other specific protein kinases. These events in turn typically lead to the modification of proteins that control gene transcription, and the activation of defense-associated gene expression.

Transcription factors and the hormone response

Numerous genes and/or proteins as well as other molecules have been identified that mediate plant defense signal transduction. Cytoskeleton and vesicle trafficking dynamics help to orient plant defense responses toward the point of pathogen attack.

Mechanisms of transcription factors and hormones

Plant immune system activity is regulated in part by signaling hormones such as:

- Salicylic acid
- Jasmonic acid
- Ethylene

There can be substantial cross-talk among these pathways.

Regulation by degradation

As with many signal transduction pathways, plant gene expression during immune responses can be regulated by degradation. This often occurs when hormone binding to hormone receptors stimulates ubiquitin-associated degradation of repressor proteins that block expression of certain genes. The net result is hormone-activated gene expression. Examples:

- Auxin: binds to receptors that then recruit and degrade repressors of transcriptional activators that stimulate auxin-specific gene expression.
- Jasmonic acid: similar to auxin, except with jasmonate receptors impacting jasmonate-response signaling mediators such as JAZ proteins.
- Gibberellic acid: Gibberellin causes receptor conformational changes and binding and degradation of Della proteins.
- Ethylene: Inhibitory phosphorylation of the EIN2 ethylene response activator is blocked by ethylene binding. When this phosphorylation is reduced, EIN2 protein is cleaved and a portion of the protein moves to the nucleus to activate ethylene-response gene expression.

Ubiquitin and E3 signaling

Ubiquitination plays a central role in cell signaling that regulates processes including protein degradation and immunological response. Although one of the main functions of ubiquitin is to target proteins for destruction, it is also useful in signaling pathways, hormone release, apoptosis and translocation of materials throughout the cell. Ubiquitination is a component of several immune responses. Without ubiquitin's proper functioning, the invasion of pathogens and other harmful molecules would increase dramatically due to weakened immune defenses.

E3 signaling

The E3 Ubiquitin ligase enzyme is a main component that provides specificity in protein degradation pathways, including immune signaling pathways. The E3 enzyme components can be grouped by which domains they contain and include several types. These include the Ring and U-box single subunit, HECT, and CRLs. Plant signaling pathways including immune responses are controlled by several feedback pathways, which often include negative feedback; and they can be regulated by De-ubiquitination enzymes, degradation of transcription factors and the degradation of negative regulators of transcription.

Plant breeding for disease resistance

Plant breeders emphasize selection and development of disease-resistant plant lines. Plant diseases can also be partially controlled by use of pesticides and by cultivation practices such as crop rotation, tillage, planting density, disease-free seeds and cleaning of equipment, but plant varieties with inherent (genetically determined) disease resistance are generally preferred. Breeding for disease resistance began when plants were first domesticated. Breeding efforts continue because pathogen populations are under selection pressure for increased virulence, new pathogens appear, evolving cultivation practices and changing climate can reduce resistance and/or strengthen pathogens, and plant breeding for other traits can disrupt prior resistance. A plant line with acceptable resistance against one pathogen may lack resistance against others.

Breeding for resistance typically includes:

- Identification of plants that may be less desirable in other ways, but which carry a useful disease resistance trait, including wild strains that often express enhanced resistance.
- Crossing of a desirable but disease-susceptible variety to another variety that is a source of resistance.
- Growth of breeding candidates in a disease-conducive setting, possibly including pathogen inoculation. Attention must be paid to the specific pathogen isolates, to address variability within a single pathogen species.
- Selection of disease-resistant individuals that retain other desirable traits such as yield, quality and including other disease resistance traits.

Resistance is termed durable if it continues to be effective over multiple years of widespread use as pathogen populations evolve. “Vertical resistance” is specific to certain races or strains of a pathogen species, is often controlled by single R genes and can be less durable. Horizontal or broad-spectrum resistance against an entire pathogen species is often only incompletely effective, but more durable, and is often controlled by many genes that segregate in breeding populations.

Crops such as potato, apple, banana and sugarcane are often propagated by vegetative reproduction to preserve highly desirable plant varieties, because for these species, outcrossing seriously disrupts the preferred traits. See also asexual propagation. Vegetatively propagated crops may be among the best targets for resistance improvement by the biotechnology method of plant transformation to manage genes that affect disease resistance.

Scientific breeding for disease resistance originated with Sir Rowland Biffen, who identified a single recessive gene for resistance to wheat yellow rust. Nearly every crop was then bred to include disease resistance (R) genes, many by introgression from compatible wild relatives.

GM or transgenic engineered disease resistance

The term GM ("genetically modified") is often used as a synonym of transgenic to refer to plants modified using recombinant DNA technologies. Plants with transgenic/GM disease resistance against insect pests have been extremely successful as commercial products, especially in maize and cotton, and are planted annually on over 20 million hectares in over 20 countries worldwide. Transgenic plant disease resistance against microbial pathogens was first demonstrated in 1986. Expression of viral coat protein gene sequences conferred virus resistance via small RNAs. This proved to be a widely applicable mechanism for inhibiting viral replication. Combining coat protein genes from three different viruses, scientists developed squash hybrids with field-validated, multiviral resistance. Similar levels of resistance to this variety of viruses had not been achieved by conventional breeding.

A similar strategy was deployed to combat papaya ringspot virus, which by 1994 threatened to destroy Hawaii's papaya industry. Field trials demonstrated excellent efficacy and high fruit quality. By 1998 the first transgenic virus-resistant papaya was approved for sale. Disease resistance has been durable for over 15 years. Transgenic papaya accounts for ~85% of Hawaiian production. The fruit is approved for sale in the U.S., Canada and Japan.

Potato lines expressing viral replicase sequences that confer resistance to potato leafroll virus were sold under the trade names NewLeaf Y and NewLeaf Plus, and were widely accepted in commercial production in 1999-2001, until McDonald's Corp. decided not to purchase GM potatoes and Monsanto decided to close their NatureMark potato business. NewLeaf Y and NewLeaf Plus potatoes carried two GM traits, as they also expressed Bt-mediated resistance to Colorado potato beetle.

No other crop with engineered disease resistance against microbial pathogens had reached the market by 2013, although more than a dozen were in some state of development and testing.

PRR transfer

Research aimed at engineered resistance follows multiple strategies. One is to transfer useful PRRs into species that lack them. Identification of functional PRRs and their transfer to a recipient species that lacks an orthologous receptor could provide a general pathway to additional broadened PRR repertoires. For example, the Arabidopsis PRR EF-Tu receptor (EFR) recognizes the bacterial translation elongation factor EF-Tu. Research performed at Sainsbury Laboratory demonstrated that deployment of EFR into either Nicotiana benthamianaor

Solanum lycopersicum (tomato), which cannot recognize EF-Tu, conferred resistance to a wide range of bacterial pathogens. EFR expression in tomato was especially effective against the widespread and devastating soil bacterium Ralstonia solanacearum. Conversely, the tomato PRR Verticillium 1 (Ve1) gene can be transferred from tomato to Arabidopsis, where it confers resistance to race 1 Verticillium isolates.

Stacking

The second strategy attempts to deploy multiple NLR genes simultaneously, a breeding strategy known as stacking. Cultivars generated by either DNA-assisted molecular breeding or gene transfer will likely display more durable resistance, because pathogens would have to mutate multiple effector genes. DNA sequencing allows researchers to functionally "mine" NLR genes from multiple species/strains.

The avrBs2 effector gene from Xanthomona perforans is the causal agent of bacterial spot disease of pepper and tomato. The first "effector-rationalized" search for a potentially durable R gene followed the finding that avrBs2 is found in most disease-causing Xanthomonas species and is required for pathogen fitness. The Bs2 NLR gene from the wild pepper, Capsicum chacoense, was moved into tomato, where it inhibited pathogen growth. Field trials demonstrated robust resistance without bactericidal chemicals. However, rare strains of Xanthomonas overcame Bs2-mediated resistance in pepper by acquisition of avrBs2 mutations that avoid recognition but retain virulence. Stacking R genes that each recognize a different core effector could delay or prevent adaptation.

More than 50 loci in wheat strains confer disease resistance against wheat stem, leaf and yellow stripe rust pathogens. The Stem rust 35 (Sr35) NLR gene, cloned from a diploid relative of cultivated wheat, Triticum monococcum, provides resistance to wheat rust isolate Ug99. Similarly, Sr33, from the wheat relative Aegilops tauschii, encodes a wheat ortholog to barley Mla powdery mildew–resistance genes. Both genes are unusual in wheat and its relatives. Combined with the Sr2 gene that acts additively with at least Sr33, they could provide durable disease resistance to Ug99 and its derivatives.

Executor genes

Another class of plant disease resistance genes opens a "trap door" that quickly kills invaded cells, stopping pathogen proliferation. Xanthomonas and Ralstonia transcription activator–like (TAL) effectors are DNA-binding proteins that activate host gene expression to enhance pathogen virulence. Both the rice and pepper lineages independently evolved TAL-effector binding sites that instead act as an executioner that induces hypersensitive host cell death when up-regulated. Xa27 from rice and Bs3 and Bs4c from pepper, are such "executor" (or "executioner") genes that encode non-homologous plant proteins of unknown function. Executor genes are expressed only in the presence of a specific TAL effector.

Engineered executor genes were demonstrated by successfully redesigning the pepper Bs3 promoter to contain two additional binding sites for TAL effectors from disparate pathogen strains. Subsequently, an engineered executor gene was deployed in rice by adding five TAL effector binding sites to the Xa27 promoter. The synthetic Xa27 construct conferred resistance against Xanthomonas bacterial blight and bacterial leaf streak species.

Host susceptibility alleles

Most plant pathogens reprogram host gene expression patterns to directly benefit the pathogen. Reprogrammed genes required for pathogen survival and proliferation can be thought of as "disease-susceptibility genes." Recessive resistance genes are disease-susceptibility candidates. For example, a mutation disabled an Arabidopsis gene encoding pectate lyase (involved in cell wall degradation), conferring resistance to the powdery mildew pathogen Golovinomyces cichoracearum. Similarly, the Barley MLO gene and spontaneously mutated pea and tomato MLO orthologs also confer powdery mildew resistance.

Lr34 is a gene that provides partial resistance to leaf and yellow rusts and powdery mildew in wheat. Lr34 encodes an adenosine triphosphate (ATP)–binding cassette (ABC) transporter. The dominant allele that provides disease resistance was recently found in cultivated wheat (not in wild strains) and, like MLO provides broad-spectrum resistance in barley.

Natural alleles of host translation elongation initiation factors eif4e and eif4g are also recessive viral-resistance genes. Some have been deployed to control potyviruses in barley, rice, tomato, pepper, pea, lettuce and melon. The discovery prompted a successful mutant screen for chemically induced eif4e alleles in tomato.

Natural promoter variation can lead to the evolution of recessive disease-resistance alleles. For example, the recessive resistance gene xa13 in rice is an allele of Os-8N3. Os-8N3 is transcriptionally activated byXanthomonas oryzae pv. oryzae strains that express the TAL effector PthXo1. The xa13 gene has a mutated effector-binding element in its promoter that eliminates PthXo1 binding and renders these lines resistant to strains that rely on PthXo1. This finding also demonstrated that Os-8N3 is required for susceptibility.

Xa13/Os-8N3 is required for pollen development, showing that such mutant alleles can be problematic should the disease-susceptibility phenotype alter function in other processes. However, mutations in the Os11N3 (OsSWEET14) TAL effector–binding element were made by fusing TAL effectors to nucleases (TALENs). Genome-edited rice plants with altered Os11N3 binding sites remained resistant to Xanthomonas oryzae pv. oryzae, but still provided normal development function.

Gene silencing

RNA silencing-based resistance is a powerful tool for engineering resistant crops. The advantage of RNAi as a novel gene therapy against fungal, viral and bacterial infection in plants lies in the fact that it regulates gene expression via messenger RNA degradation, translation repression and chromatin remodelling through small non-coding RNAs. Mechanistically, the silencing processes are guided by processing products of the double-stranded RNA (dsRNA) trigger, which are known as small interfering RNAs and microRNAs.

Host range

Among the thousands of species of plant pathogenic microorganisms, only a small minority have the capacity to infect a broad range of plant species. Most pathogens instead exhibit a high degree of host-specificity. Non-host plant species are often said to express non-host resistance. The term host resistance is used when a pathogen species can be pathogenic on the host species but certain strains of that plant species resist certain strains of the pathogen species. The causes of host resistance and non-host resistance can overlap. Pathogen host range is determined, among other things, by the presence of appropriate effectors that allow colonization of a particular host. Pathogen host range can change quite suddenly if, for example, the pathogen's capacity to synthesize a host-specific toxin or effector is gained by gene shuffling/mutation, or by horizontal gene transfer.

EPIDEMICS AND POPULATION BIOLOGY

Native populations are often characterized by substantial genotype diversity and dispersed populations (growth in a mixture with many other plant species). They also have undergone of plant-pathogen coevolution. Hence as long as novel pathogens are not introduced/do not evolve, such populations generally exhibit only a low incidence of severe disease epidemics.

Monocrop agricultural systems provide an ideal environment for pathogen evolution, because they offer a high density of target specimens with similar/identical genotypes.The rise in mobility stemming from modern transportation systems provides pathogens with access to more potential targets. Climate change can alter the viable geographic range of pathogen species and cause some diseases to become a problem in areas where the disease was previously less important.

These factors make modern agriculture more prone to disease epidemics. Common solutions include constant breeding for disease resistance, use of pesticides, use of border inspections and plant import restrictions, maintenance of significant genetic diversity within the crop gene pool, and constant surveillance to accelerate initiation of appropriate responses. Some pathogen species have much greater capacity to overcome plant disease resistance than others, often because of their ability to evolve rapidly and to disperse broadly.

9

HORTICULTURAL ENGINEERING

Horticultural engineering blends agricultural engineering, plant science, computer science, and control theory to produce effective and efficient plant growing systems, whether in high-tech greenhouses or low-tech row cover systems. In the case of greenhouse production systems, another frequently used term is Controlled Environment Agriculture. People trained in this discipline have a strong understanding of and appreciation for the engineering, biological, and environmental aspects of plant production. Horticultural engineers play a vital role in the secure, healthy, and cost-effective production of many edible and ornamental plant species on which we all depend. Agricultural engineering is the engineering discipline that applies engineering science and technology to agricultural production and processing. Agricultural engineering combines the disciplines of animal biology, plant biology, and mechanical, civil and chemical engineering principles with a knowledge of agricultural principles.

SUBFIELDS

Some of the specialties of agricultural engineers include:

- the design of agricultural machinery, equipment, and agricultural structures
- crop production, including seeding, tillage, irrigation and the conservation of soil and water
- animal production, including the care and processing of poultry and fish and dairy management

- the processing of food and other agricultural and biorenewable products, and food engineering.

Agricultural Engineers

Agricultural Engineers may perform tasks as planning, supervising and managing the building of dairy effluent schemes, irrigation, drainage, flood and water control systems, perform environmental impact assessments, agricultural product processing and interpret research results and implement relevant practices. A large percentage of agricultural engineers work in academia or for government agencies such as the United States Department of Agriculture or state agricultural extension services. Many are employed by manufacturers of agricultural machinery, equipment and agricultural product processing. Agricultural engineers work in production, sales, management, research and development, or applied science.

Agricultural Machinery

Agricultural machinery is one of the most revolutionary and impactful applications of modern technology. The truly elemental human need for food has often driven the development of technology and machines. Over the last 250 years, advances in farm equipment have transformed the way people are employed and produce their food worldwide.

History

Hand Tools

The first person to turn from the hunting and gathering lifestyle to farming probably did so by using his bare hands, and perhaps some sticks or stones. Tools such as knives, scythes, and wooden plows were eventually developed, and dominated agriculture for thousands of years. During this time, almost everyone worked in agriculture, because each family could barely raise enough food for themselves with the limited technology of the day.

The Industrial Revolution

With the coming of the Industrial Revolution and the development of more complicated machines, farming methods took a great leap forward. Instead of harvesting grain by hand with a sharp blade, wheeled machines cut a continuous swath. Instead of threshing the grain by beating it with sticks, threshing machines separated the seeds from the heads and stalks.

Steam Power

These machines required a lot of power, which was originally supplied by horses or other domesticated animals. With the invention of steam power came the portable engine and later the traction engine, a multipurpose, mobile energy source that was the ground-crawling cousin to the steam locomotive. Agricultural steam engines took over the heavy pulling work of horses, and were also equipped with a pulley that could power stationary machines via the use of a long belt. The steam-powered behemoths were low-powered by today's standards but, because of their size and their low gear ratios, they could provide a large drawbar pull. Their slow speed led farmers to comment that tractors had two speeds: "slow, and darn slow."

The Internal Combustion Engine

Gasoline, and later diesel engines became the main source of power for the next generation of tractors. These engines also contributed to the development of the self-propelled, combined harvester and thresher, or combine for short. Instead of cutting the grain stalks and transporting them to a stationary threshing machine, these combines cut, threshed, and separated the grain while moving continuously through the field.

Types

Combines might have taken the harvesting job away from tractors, but tractors still do the majority of work on a modern farm. They are used to pull implements—machines that till the ground, plant seed, or perform a number of other tasks.

Tillage implements prepare the soil for planting by loosening the soil and killing weeds or competing plants. The best-known is the plow, the ancient implement that was upgraded in 1838 by a man named John Deere. Plows are actually used less frequently in the U.S., today, with offset disks used instead to turn over the soil and chisels used to gain the depth needed to retain moisture.

The most common type of seeder is called a planter and spaces seeds out equally in long rows, which are usually 2 to 3 feet apart. Some crops are planted by drills, which put out much more seed in rows less than a foot apart, blanketing the field with crops. Transplanters fully or partially automate the task of transplanting seedlings to the field. With the widespread use of plastic mulch, plastic mulch layers, transplanters, and seeders lay down long rows of plastic, and plant through them automatically.

After planting, other implements can be used to cultivate weeds from between rows, or to spread fertilizer and pesticides. Hay balers can be used to tightly package grass or alfalfa into a storable form for the winter months.

Modern irrigation also relies on a great deal of machinery. A variety of engines, pumps and other specialized gear are used to provide water quickly and in high volumes to large areas of land. Similar types of equipment can be used to deliver fertilizers and pesticides.

And, besides the tractor, a variety of vehicles have been adapted for use in various aspects of farming, including trucks, airplanes, and helicopters, for everything from transporting crops and making equipment mobile, to aerial spraying and livestock herd management.

New Technology and the Future

The basic technology of agricultural machines has changed little in the last century. Though modern harvesters and planters may do a better job or be slightly tweaked from their predecessors, the US$250,000 combine of today still cuts, threshes, and separates grain in essentially the same way it has always been done. However, technology is changing the way that humans operate the machines, as computer monitoring systems, GPS locators, and self-steer programs allow the most advanced tractors and implements to be more precise and less wasteful in the use of fuel, seed, or fertilizer. In the foreseeable future, some agricultural machines will be capable of driving themselves, using GPS maps and electronic sensors. Even more esoteric are the new areas of nanotechnology and genetic engineering, where submicroscopic devices and biological processes, respectively, are being used as machines to perform agricultural tasks in unusual new ways.

It is estimated that at the turn of the 20th century, one farmer in the U.S. could feed 25 people, where today, that ratio is 1:130 (in a modern grain farm, a single farmer can produce cereal to feed over a thousand people). With continuing advances in agricultural machinery, the role of the farmer will become increasingly specialized and rare.

Agricultural Machinery

Agricultural machinery is any kind of machinery used on a farm to help with farming. The best-known example of this kind is the tractor.

Traction and power:

- Tractor
- Crawler tractor / Caterpillar tractor

Soil cultivation:

- Cultivator
- Cultipacker

- Chisel plow
- Harrow
    - o Spike harrow
    - o Drag harrow
    - o Disk harrow
- Plough
- Power tiller / Rotary tiller / Rototiller
- Spading machine
- Subsoiler
- Walking tractor
- Stone Picker, also known as a Rock Picker

Planting:

- Broadcast seeder (alternately: broadcast spreader or fertilizer spreader)
- Planter (farm implement)
- Plastic mulch layer
- Potato planter
- Seed drill
- Air seeder
- Precision drill
- Transplanter
    - o Rice transplanter

Fertilizing & Pest Control:

- Fertilizer spreader,
- Terragator
- Manure spreader

Irrigation:

- Center pivot irrigation

Harvesting / post-harvest:

- beet harvester

- Bean harvester
- Cane Harvester
- Chaser Bin
- Combine harvester
- Conveyor belt
- Corn harvester
- Cotton picker
- Fanning mill
- Forage harvester (or silage harvester)
- Gleaner
- Gravity wagon
- Haulout Transporter
- Huller
- Over-the-row mechanical harvester for harvesting apples
- Potato digger
- Potato harvester
- Sickle
- Swather

Hay making:

- Bale mover
- Baler and Big Baler
- Conditioner
- Hay rake
- Hay tedder
- Mower
- Self-loading wagon - used in Europe, but not common in USA

Loading:

- Backhoe
- Front end loader

- Skid-steer loader

Other:

- Allen Scythe
- Grain auger
- Feed grinder
- Grain cart
- Rock picker
- Conveyor Analyzer

Obsolete farm machinery:

Steam-powered:

- Stationary steam engine
- Portable engine
- Traction engine
    - o Agricultural engine
    - o Ploughing engine
    - o Steam tractor
- Binder
- Hog oiler
- Reaper
- Threshing machine
- Drag harrow

BIOPROCESS ENGINEERING

Bioprocess Engineering is a specialization of Chemical Engineering or of Agricultural Engineering. It deals with the design and development of equipment and processes for the manufacturing of products such as food, feed, pharmaceuticals, nutraceuticals, chemicals, and polymers and paper from biological materials.

Food Engineering

Food engineering is a multidisciplinary program which combines science, microbiology, and engineering education for food and related industries. Food engineering includes, but is not limited to, the application of agricultural engineering and chemical engineering principles to food materials. Food engineering is a very wide field of activities. Prospective major employers for food engineers include companies involved in food processing, food machinery, packaging, ingredient manufacturing, instrumentation and control. Firms that design and build food processing plants, consulting firms, government agencies, pharmaceutical, and health-care firms also hire food engineers. Among its domain of knowledge and action are:

- Research and development of new foods, biological and pharmaceutical products
- Development and operation of manufacturing, packaging and distributing systems for drug/food products
- Design and installation of food/biological/pharmaceutical production processes
- Design and operation of environmentally responsible waste treatment systems
- Marketing and technical support for manufacturing plants.

Energy

In physics, energy is a scalar physical quantity that describes the amount of work that can be performed by a force, an attribute of objects and systems that is subject to a conservation law. Eight different forms of energy exist to explain all known natural phenomena. These forms include (but are not limited to) kinetic, potential, thermal, gravitational, sound, light, elastic, and electromagnetic energy. The forms of energy are often named after a related force.

Any form of energy can be transformed into another form, but the total energy always remains the same. This principle, the conservation of energy, was first postulated in the early 19th century, and applies to any isolated system. According to Noether's theorem, the conservation of energy is a consequence of the fact that the laws of physics do not change over time.

Although the total energy of a system does not change with time, its value may depend on the frame of reference. For example, a seated passenger in a moving airplane has zero kinetic energy relative to the airplane, but non-zero kinetic energy relative to the Earth.

10

PEST MANAGEMENT

People are health-conscious and they eat healthy, pest-free food in their regular diet. Pest management relies on a combination of common-sense practices. This management includes a broad range of programs concerning human health, environmental and economic issues related to the management of pest populations through a variety of science-based technologies. Nowadays, people want safe, pest and disease-free homes, schools, parks, recreational areas, as well as a safe and affordable supply of blemish-free food products and a wholesome pesticide-free environment.

Biological control can be defined as the deliberate use of natural enemies—predators, parasites, pathogens, and competitors to suppress and maintain populations of a target pest species (insects, mites, weeds, plant pathogens, and other pest organisms). NIFA supports research in bio-based pest management, which has the goal of providing safer and more effective methods of controlling pests while reducing our reliance on synthetic pesticides.

The organic industry is the fastest growing segment of U.S. agriculture, with sales growing from $1 billion in 1990 to a projected $23.9 billion in 2009. NIFA supports programs to address critical organic agriculture issues, priorities, or problems through the integration of research and extension activities and programs to evaluate both the environmental impacts of organic agriculture and the environmental services provided.

Plant breeding, genetics, and genomics have a critical role to play in 21st century. Through initiatives, projects, and funding programs, NIFA provides leadership and coordination for national genomics research, supports plant breeding through a multi-state effort to identify critical needs, and helps to focus federal dollars for research on cutting-edge plant improvement for human and animal health, food security, sustainable production, and bioenergy.

INTEGRATED PEST MANAGEMENT

NIFA funds programs and projects which support research, education, and extension activities that promote pest management in general, and reduced risk pest management in particular. The agency's pest management programs are implemented through working partnerships with scientists in nation's colleges and universities, other federal agencies and the private sector.

Integrated pest management is an integrated approach of crop management to solve ecological problems when applied in agriculture. These methods are performed in three stages: prevention, observation, and intervention. It is an ecological approach with a main goal of significantly reducing or eliminating the use of pesticides while at the same time managing pest populations at an acceptable level. For their leadership in developing and spreading IPM worldwide.

In the second half of the last century, when synthetic insecticides became widely available, entomologists in California developed the concept of "supervised insect control". Around the same time, entomologists in cotton-belt states such as Arkansas were advocating a similar approach. Under this scheme, insect control was "supervised" by qualified entomologists, and insecticide applications were based on conclusions reached from periodic monitoring of pest and natural-enemy populations. This was viewed as an alternative to calendar-based insecticide programs. Supervised control was based on a sound knowledge of the ecology and analysis of projected trends in pest and natural-enemy populations.

Supervised control formed much of the conceptual basis for the "integrated control" that University of California entomologists articulated in the 1950s. Integrated control sought to identify the best mix of chemical and biological controls for a given insect pest. Chemical insecticides were to be used in manner least disruptive to biological control. The term "integrated" was thus synonymous with "compatible." Chemical controls were to be applied only after regular monitoring indicated that a pest population had reached a level (the economic threshold) that required treatment to prevent the population from reaching a level (the economic injury level) at which economic losses would exceed the cost of the artificial control measures.

IPM extended the concept of integrated control to all classes of pests and was expanded to include tactics other than just chemical and biological controls. Artificial controls such as pesticides were to be applied as in integrated control, but these now had to be compatible with control tactics for all classes of pests. Other tactics, such as host-plant resistance and cultural manipulations, became part of the IPM arsenal. IPM added the multidisciplinary element, involving entomologists, plant pathologists, nematologists, and weed scientists.

In the United States, IPM was formulated into national policy in February 1972 when President Nixon directed federal agencies to take steps to advance the concept and application of IPM in all relevant sectors. In 1979, President Carter established an interagency IPM Coordinating Committee to ensure development and implementation of IPM practices. (references: "The History of IPM", BioControl Reference Center.

An IPM system is designed around six basic components: The US Environmental Protection Agency has a useful set of IPM principles.

1. Acceptable pest levels: The emphasis is on control, not eradication. IPM holds that wiping out an entire pest population is literally impossible, and the attempt can be econmically expensive, environmentally unsafe, and frequently unachievable. IPM programmes first work to establish acceptable pest levels, called action thresholds, and apply controls if those thresholds are crossed. These thresholds are pest and site specific, meaning that it may be acceptable at one site to have a weed such as white clover, but at another site it may not be acceptable. By allowing a pest population to survive at a reasonable threshold, selection pressure is reduced. This stops the pest gaining resistance to chemicals produced by the plant or applied to the crops. If many of the pests are killed then any that have resistance to the chemical will form the genetic basis of the future, more resistant, population. By not killing all the pests there are some unresistant pests left that will dilute any resistant genes that appear.
2. Safe agricultural practices: Selecting varieties best for local growing conditions, and maintaining healthy crops, is the initial step of defense, together with plant quarantine and 'cultural techniques' such as crop sanitation (e.g. removal of diseased plants to prevent spread of infection).
3. Regular observation : It is the cornerstone of IPM. Observation is broken into two steps, first; inspection and second; identification. Visual inspection, insect and spore traps, and other measurement methods and monitoring tools are used to monitor pest levels. Accurate pest identification is critical to a successful

IPM program. Record-keeping is essential, as is a thorough knowledge of the behavior and reproductive cycles of target pests. Since insects are cold-blooded, their physical development is dependent on the temperature of their environment. Many insects have had their development cycles modeled in terms of degree days. Monitor the degree days of an environment to determine when is the optimal time for a specific insect's outbreak.

4. Mechanical controls: Should a pest reach an unacceptable level, mechanical methods are the first options to consider. They include simple hand-picking, erecting insect barriers, using traps, vacuuming, and tillage to disrupt breeding.
5. Biological controls: Natural biological processes and materials can provide control, with minimal environmental impact, and often at low cost. The main focus here is on promoting beneficial insects that eat target pests. Biological insecticides, derived from naturally occurring microorganisms (For example, Bt, entomopathogenic fungi and entomo-pathogenic nematodes), also fit in this category.
6. Limited Pesticide Use: Synthetic pesticides are generally only used as required and often only at specific times in a pests life cycle. Many of the newer pesticide groups are derived from plants or naturally occurring substances (e.g.: nicotine, pyrethrum and insect juvenile hormone analogues), but the toxophore or active component may be altered to provide increased biological activity or stability. Further 'biology-based' or 'ecological' techniques are under evaluation.

An Integrated pest management regime can be quite simple or sophisticated. Historically, the main focus of IPM programmes was on agricultural insect pests. Although originally developed for agricultural pest management, IPM programmes are now developed to encompass diseases, weeds, and other pests that interfere with the management objectives of sites such as residential and commercial structures, lawn and turf areas, and home and community gardens.

Integrated pest management is applicable to all types of agriculture and sites such as residential and commercial structures, lawn and turf areas, and home and community gardens. Reliance on knowledge, experience, observation, and integration of multiple techniques makes IPM a perfect fit for organic farming (sans artificial pesticide application). For large-scale, chemical-based farms, IPM can reduce human and environmental exposure to hazardous chemicals, and potentially lower overall costs of pesticide application material and labor.

Cases of mistaken identity may result in ineffective actions. If plant damage due to over-watering are mistaken for fungal infection, spray costs can be incurred, and the plant is no better off.

Learn pest and host life cycle and biology. At the time you see a pest, it may be too late to do much about it except maybe spray with a pesticide . Often, there is another stage of the life cycle that is susceptible to preventative actions. For example, weeds reproducing from last year's seed can be prevented with mulches. Also, learning what a pest needs to survive allows you to remove these.

Preventative Actions

Monitor or sample environment for pest population - How many are here? Preventative actions must be taken at the correct time if they are to be effective. For this reason, once the pest is correctly identified, monitoring must begin before it becomes a problem. For example, in school cafeterias where roaches may be expected to appear, sticky traps are set out before school starts. Traps are checked at regular intervals so populations can be monitored and controlled before they get out of hand. Some factors to consider and monitor include: Is the pest present/ absent? What is the distribution - all over or only in certain spots? Is the pest population increasing or decreasing?

In some cases, a certain number of pests can be tolerated. Soybeans are quite tolerant of defoliation, so if there are a few caterpillars in the field and their population is not increasing dramatically, there is not necessarily any action necessary. Conversely, there is a point at which action must be taken to control cost. For the farmer, that point is the one at which the cost of damage by the pest is more than the cost of control. This is an economic threshold. Tolerance of pests varies also by whether or not they are a health hazard (low tolerance) or merely a cosmetic damage (high tolerance in a non-commercial situation).

Different sites may also have varying requirements based on specific areas. White clover may be perfectly acceptable on the sides of a tee box on a golf course, but unacceptable in the fairway where it could cause confusion in the field of play.

Choose an appropriate combination of management tactics For any pest situation, there will be several options to consider. Options include, mechanical or physical control, cultural controls, biological controls and chemical controls. Mechanical or physical controls include picking pests off plants, or using netting or other material to exclude pests such as birds from grapes or rodents from structures. Cultural controls include keeping an area free of conducive conditions by removing or storing waste properly, removing diseased areas of plants properly. Biological controls can be support either through conservation of natural predators or augmentation of natural predators.

Augmentative control includes the introduction of naturally occurring predators at either an inundative or inoculative level. An inundative release would be one that seeks to inundate a site with a pest's predator to impact the pest population. An inoculative release would be a smaller number of pest predators to supplement the natural population and provide ongoing control. Chemical controls would include horticultural oils or the application of pesticides such as insecticides and herbicides. A Green Pest Management program would use pesticides derived from plants, such as botanicals, or other naturally occurring materials.

Evaluation is often one of the most important steps. This is the process to review an IPM program and the results it generated. Asking the following questions is useful: Did actions have the desired effect? Was the pest prevented or managed to farmer satisfaction? Was the method itself satisfactory? Were there any unintended side effects? What can be done in the future for this pest situation? Understanding the effectiveness of the IPM program allows the site manager to make modifications to the IPM plan prior to pests reaching the action threshold and requiring action again.

What is Pest Control?

It is the regulation or management of a species defined as a pest, usually because it is perceived to be detrimental to a person's health, the ecology or the economy.

Pest control is at least as old as agriculture, as there has always been a need to keep crops free from pests. In order to maximize food production, it is advantageous to protect crops from competing species of plants, as well as from herbivores competing with humans.

The conventional approach was probably the first to be employed, since it is comparatively easy to destroy weeds by burning them or plowing them under, and to kill larger competing herbivores, such as crows and other birds eating seeds. Techniques such as crop rotation, companion planting (also known as intercropping or mixed cropping), and the selective breeding of pest-resistant cultivars have a long history.

In the United Kingdom, following concern about animal welfare, human pest control and deterrence is gaining ground through the use of animal psychology rather than destruction. For instance, with the urban Red Fox which territorial behaviour is used against the animal, usually in conjunction with non-injurious chemical repellents. In rural areas of Britain, the use of firearms for pest control is quite common. Airguns are particularly popular for control of small pests such as rats, rabbits and grey squirrels, because of their lower power they can be used in more restrictive spaces such as gardens, where using a firearm would be unsafe.

Chemical pesticides date back 4,500 years, when the Sumerians used sulfur compounds as insecticides. The Rigveda, which is about 4,000 years old, also mentions the use of poisonous plants for pest control. It was only with the industrialization and mechanization of agriculture in the 18th and 19th century, and the introduction of the insecticides pyrethrum and derris that chemical pest control became widespread. In the 20th century, the discovery of several synthetic insecticides, such as DDT, and herbicides boosted this development. Chemical pest control is still the predominant type of pest control today, although its long-term effects led to a renewed interest in traditional and biological pest control towards the end of the 20th century.

Causes

Many pests have only become a problem because of the direct actions of humans. Modifying these actions can often substantially reduce the pest problem. In the United States, raccoons caused a nuisance by tearing open refuse sacks. Many householders introduced bins with locking lids, which deterred the raccoons from visiting.

House flies tend to accumulate wherever there is human activity and is virtually a global phenomenon, especially where food or food waste is exposed. Similarly, seagulls have become pests at many seaside resorts. Tourists would often feed the birds with scraps of fish and chips, and before long, the birds would become dependent on this food source and act aggressively towards humans. Living organisms evolve and increase their resistance to biological, chemical, physical or any other form of control. Unless the target population is completely exterminated or is rendered incapable of reproduction, the surviving population will inevitably acquire a tolerance of whatever pressures are brought to bear—this results in an evolutionary arms race.

Types of pest control

The best and most effective way for controlling pests is pest management which we have already discussed. The first and important step in pest control is to identify the pest problem. We can control pests by many means, which we will discuss here one by one.

Biological Pest Control

The point of biological pest control, or any natural pest control, is to eliminate a pest with minimal harm to the ecological balance of the environment in its present form. Biological pest control is the control of one through the control and management of natural predators and parasites. For example: mosquitoes are often controlled by putting Bt Bacillus thuringiensis spp. israelensis, a bacterium that infects and kills mosquito larvae, in local water sources. The treatment has no known negative consequences on the remaining ecology and is safe for humans to drink.

Elimination of Breeding Grounds

Proper waste management and drainage of still water, eliminates the breeding ground of many pests.

Garbage provides food and shelter for many unwanted organisms, as well as an area where still water might collect and be used as a breeding ground by mosquitoes. Communities that have proper garbage collection and disposal, have far less of a problem with rats, cockroaches, mosquitoes, flies and other pests than those that don't.

Open air sewers are ample breeding ground for various pests as well. By building and maintaining a proper sewer system, this problem is eliminated.

Poisoned Meats

It is a common method for controlling rat populations, however is not as effective when there are other food sources around, such as garbage. Poisoned meats have been used for centuries for killing off wolves, birds that were seen to threaten crops, and against other creatures.

Field Burning

Traditionally, after a sugar cane harvest, the fields are all burned, to kill off any insects or eggs that might be in the fields.

Hunting

Historically, in some European countries, when stray dogs and cats became too numerous, local populations gathered together to round up all animals that did not appear to have an owner and kill them. In some nations, teams of rat catchers work at chasing rats from the field, and killing them with dogs and simple hand tools. Some communities have in the past employed a bounty system, where a town clerk will pay a set fee for every rat head brought in as proof of a rat killing.

Traps

Traps have been used for killing off mice found in houses, for killing wolves, and for capturing raccoons and stray cats and dogs for disposal by town officials.

POISON SPRAY

Spraying poisons by planes, hand held units, or trucks that carry the spraying equipment, is a common method of pest control. Throughout the United States of America, towns often drive a town owned truck around once or twice a week to each street, spraying for mosquitoes. Crop dusters commonly fly over farmland and spray poison to kill off pest that would threaten the crops. Many find spraying poison around their yard, homes, or businesses, far more desirable than allowing insects to thrive there.

Space Fumigation

A project that involves a structure be covered or sealed airtight followed by the introduction of a penetrating, deadly gas at a killing concentration a long period of time (24-72hrs.). Although expensive, space fumigation targets all life stages of pests.

Space Treatment

A long term project involving fogging or misting type applicators. Liquid insecticide is dispersed in the atmosphere within a structure. Treatments do not require the evacuation or airtight sealing of a building, allowing most work within the building to continue but at the cost of the penetrating effects. Contact insecticides are generally used, minimizing the long lasting residual effects. On August 10, 1973, the Federal Register printed the definition of Space treatment as defined by the U.S. Environmental Protection Agency (EPA):

- the dispersal of insecticides into the air by foggers, misters, aerosol devices or vapor dispensers for control of flying insects and exposed crawling insects“

Sterilization

Laboratory studies conducted with U-5897 (3-chloro-1,2-propanediol) were attempted in the early 1970s although these proved unsuccessful. Research into tgghhrilization bait is ongoing.

Another effective method of soil sterilization is soil steaming. Pest is killed through hot steam which is induced into the soil.

Destruction of Infected Plants

Forest services sometimes destroy all the trees in an area where some are infected with insects, if seen as necessary to prevent the insect species from spreading. Farms infested with certain insects, have been burned entirely, to prevent the pest from spreading elsewhere.

Natural Rodent Control

Several wildlife rehabilitation organizations encourage natural form of rodent control through exclusion and predator support and preventing secondary poisoning altogether.

The United States Environmental Protection Agency agrees, noting in its Proposed Risk Mitigation Decision for Nine Rodenticides that "without habitat modification to make areas less attractive to commensal rodents, even eradication will not prevent new populations from recolonizing the habitat."

Repellents

- Balsam fir oil from the tree Abies balsamea is an EPA approved non-toxic rodent repellent.
- Acacia polyacantha subsp. campylacantha root emits chemical compounds that repel animals including crocodiles, snakes and rats.

Insectary Plants

It is a term used to describe plants that attract insects. As such, beneficial insectary plants are intentionally introduced into an ecosystem to increase pollen resources and nectar resources required by the natural enemies of harmful or unwanted insect pests. Beyond an effective natural control of pests, the friendly insects also assist in pollination.

The "friendly insects" include ladybeetles, bees, ground beetles, hoverflies, and parasitic wasps. Other animals that are frequently considered beneficial include lizards, spiders, toads, and humming birds. Beneficial insects are as much as ten times more abundant in the insectary plantings area.

Mortality of scale insects (caused by natural enemies) can be double with insectary plantings. Additionally, a diversity of insectary plants can increase the population of beneficial insects such that these levels can be sustained even when the insectary plants are removed or die off.

For maximum benefit in the garden, insectary plants can be grown alongside desired garden plants that do not have this benefit. The insects attracted to the insectary plants will also help the other nearby garden plants.

Many members of the Apiaceae or Umbelliferae family are excellent insectary plants. Fennel, angelica, coriander (cilantro), dill, and wild carrot all provide in great number the tiny flowers required by parasitic wasps. Various clovers, yarrow, and rue also attract parasitic and predatory insects.

Low-growing plants, such as thyme, rosemary, or mint, provide shelter for ground beetles and other beneficial insects. Composite flowers (daisy and chamomile) and mints (spearmint, peppermint, or catnip) will attract predatory wasps, hoverflies, and robber flies. The wasps will catch caterpillars and grubs to feed their young, while the predatory and parasitic flies attack many kinds of insects, including leafhoppers and caterpillars.

Other insectary plants include: mustard (*Brassica juncea*), phacelia (Phacelia tanacetifolia), buckwheat (Fagopyrum esceulentum), marigold (Tagetes patula), elderberry (sambucus mexicana), and Korean licorice mint (Agastache rugosa).

Beneficial Weed

A beneficial weed is any of various plants not generally considered domesticated, but which nonetheless has some companion plant effect, or else is edible or somehow beneficial. Beneficial weeds include a great many wildflowers, as well as other weeds that are commonly removed or poisoned.

Although errantly assumed to compete with neighboring plants for food and moisture, some "weeds" provide the soil with nutrients, either directly or indirectly.

- Weeds with strong, widespread roots also introduce organic matter to the earth in the form of those roots, turning hard, dense clay dirt into richer, more fertile soil.
- For example, legumes, such as white clover, add nitrogen to the soil through the process of nitrogen fixation, where bacteria symbiotically living in their roots extract nitrogen from the atmosphere, making it accessible in the soil for its host, and any nearby plants.
- Others use deep tap roots to bring up nutrients and moisture from beyond the range of normal plants, so that the soil improves in quality over generations of that plant's presence.

In fact, some common plants like tomatoes and corn will "piggyback" on nearby weeds, allowing their relatively weak root systems to go deeper than they could have alone.

Pests

Many weeds protect nearby plants from insect pests.One way they can do this is to repel insects and other pests through their smell , as do alliums and wormwood. Another is to entirely mask a companion's scent, or the pheromones of pest insects, as with ground ivy and wild oregano.

Some are also unpleasant to small animals, because of their spines or other features, keeping them away from an area to be protected.

TRAP CROPS

Some weeds act as trap crops, distracting pests away from valued plants. Insects seeking a food plant search by smell, and then land at random on anything green in the area of the scent. If they land on an edible "weed", they will stay there instead of going on to the intended victim. Sometimes, they actively prefer the trap crop.

Host-finding Disruption

Recent studies on host-plant finding have shown that flying pests are far less successful if their host-plants are surrounded by any other plant or even "decoy-plants" made of green plastic, cardboard, or any other green material.

- First, they seek plants by scent. Any "weed" that has a scent reduces the odds of them finding crop plants. Obvious examples are Crow Garlic and Ground Ivy, the former being "wild chives" and the latter a form of wild mint, both dramatically masking both plant scent and insect pheromones. They cut down Japanese beetle infestation, and caterpillar infestation, for example cabbage worm, tomato hornworm, and even squash bugs.
- Second, once an insect is near its target, it avoids landing on dirt, but lands on the nearest green thing. Bare earth gardening helps them home in perfectly on the victim crop. But if one is using "green mulch", even grass or clover, the odds are that they will make what's called an "inappropriate landing" on some green thing they don't want. They will then fly a short distance at random, and land on any other green thing. If they fail to accidentally hit the right kind of plant after several tries, they give up.
- If they are there to lay eggs on the crop, weeds provide one more line of defense: Even if they find the right plant, in order to ensure that they didn't hit on a dying plant or falling leaf, they then make short leaf-to-leaf flights before laying eggs. They must land on the "right kind of leaf" enough times in sequence, before they will risk laying their eggs. The more other greenery is nearby, the harder it is for them to remain on target and get

enough reinforcement. Enough "inappropriate landings", and they give up, heading elsewhere.

One scientific study said that simply having clover growing nearby cut the odds of cabbage root flies hitting the right plant from 36% to 7%. Statistically, this is enough to dramatically change the productivity of one's crop, all alone.

Many plants can grow intercropped in the same space, because they exist on different levels in the same area, providing ground cover or working as a trellis for each other. This healthier style of horticulture is called forest gardening.

Among the obvious benefits are providing a wind break or shelter from noonday sun for more delicate plants.

GREEN MULCH

Conversely, some intercropped provide a living mulch effect, used by inhibiting the growth of any weeds that are actually harmful, and creating a humid, cooler microclimate around nearby plants, stabilizing soil moisture more than they consume it for themselves.

Plants such as ryegrass, red clover, and white clover are examples of "weeds" that are living mulches, often welcomed in horticulture.

Herbicide

Repel plants or fungi, through a chemical means known as allelopathy. Specific other plants can be bothered by a chemical emission through their roots or air, slowing their growth, preventing seed germination, or even killing them.

Beneficial Insects

A common companion plant benefit from many weeds is to attract, or be inhabited by, beneficial insects or other organisms which benefit plants.

For example, wild umbellifers attract predatory wasps and flies, that eat nectar, but reproduce by feeding common garden pests to their offspring .

Likewise, some weeds attract ladybugs or the "good" types of nematode, or provide ground cover for predatory beetles.

Human-useful

Some beneficial weeds, such as lamb's quarters, are edible. This list of edible flowers includes many wildflowers that are considered weeds when not planted intentionally. Dandelion is an excellent example of an edible weed.

A number of weeds have been proposed as natural alternate sources for latex (rubber), including goldenrod, from which the tires were made on the car famously given by Henry Ford, to Thomas Edison.

Cocklebur and stinging nettle have been used for natural dyes.

Studies have shown that milkweed is actually a more effective insulator than goose down.

Flavor

Some plants seem to subtly change the flavor of other plants around them, in a way humans find desirable, for example stinging nettle, besides being edible if properly cooked, seems to increase essential oil production in nearby herbs.

Well-known Flavor

Clover is a legume. Like other beans, it hosts bacteria that fix nitrogen in the soil. Its vining nature covers the ground, sheltering more moisture than it consumes, providing a humid, cooler microclimate for surrounding plants as a "green mulch". It also is preferred by rodents over many garden crops, cutting down on the amount of desired vegetables that get eaten.

Dandelions possess a deep, strong tap root that breaks up hard soil, benefiting weaker-rooted plants nearby, and draw up nutrients from deeper than shallower-rooted nearby plants can access. They will also excrete minerals and nitrogen through their roots.

Crow garlic, the wild chives found in sunny parts of a North American yard, has all of the companion plant benefits of other alliums, including repelling japanese beetles, aphids, and rodents, and being believed to benefit the flavor of solanums like tomatoes and peppers. It can be used as a substitute for some alliums in cooking, if its slightly more bitter flavor is masked by sweetness or lipids.

Bishop's lace or Queen Anne's Lace works as a nurse plant for nearby crops like lettuce, shading them from overly intense sunlight and keeping more humidity in the air. It attracts predatory wasps and flies that eat vegetable pests. It has a scientifically tested beneficial effect on nearby tomato plants, and when young has an edible root, both of these because the domesticated carrot is simply a cultivar of this "weed".

Pesticide application

It refers to the practical way in which pesticides, including herbicides, fungicides, insecticides, or nematode control agents are delivered to their biological targets like pest organism, crop or other plant. Public concern about the use of pesticides has high-lighted the need to make this process as efficient as possible, in order to minimise their release into the environment and human exposure including operators, bystanders and consumers of produce.

The practice of pest management by the rational application of pesticides is supremely multi-disciplinary, combining many aspects of biology and chemistry with: agronomy, engineering, meteorology, socio-economics and public health, together with newer disciplines such as biotechnology and information science.

SEED TREATMENTS AND SPRAY APPLICATION

Seed treatments can achieve exceptionally high efficiencies, in terms of effective dose-transfer to a crop. Pesticides are applied to the seed prior to planting, in the form of a seed treatment, or coating, to protect against soil-borne risks to the plant; additionally, these coatings can provide supplemental chemicals and nutrients designed to encourage growth. A typical seed coating can include a nutrient layer-containing nitrogen, phosphorus, and potassium, a rhizobial layer-containing symbiotic bacteria and other beneficial microorganisms, and a fungicide (or other chemical) layer to make the seed less vulnerable to pests.

One of the more common forms of pesticide application, especially in conventional agriculture, is the use of mechanical sprayers. Hydraulic sprayers consists of a tank, a pump, a lance (for single nozzles) or boom, and a nozzle (or multiple nozzles). Sprayers convert a pesticide formulation, often containing a mixture of water (or another liquid chemical carrier, such as fertilizer) and chemical, into droplets, which can be large rain-type drops or tiny almost-invisible particles.

This conversion is accomplished by forcing the spray mixture through a spray nozzle under pressure. The size of droplets can be altered through the use of different nozzle sizes, or by altering the pressure under which it is forced, or a combination of both. Large droplets have the advantage of being less susceptible to spray drift, but require more water per unit of land covered. Due to static electricity, small droplets are able to maximize contact with a target organism, but very still wind conditions are required.

Pre-emergent Application

Post-emergent pesticide application requires the use of specific chemicals chosen minimize harm to the desirable target organism. An example is 2,4-Dichlorophenoxyacetic acid, which will injure broadleaf weeds (dicots) but leave behind grasses (monocots). Such a chemical has been used extensively on wheat crops, for example. A number of companies have also created genetically-modified organisms that are resistant to various pesticides. Examples include glyphosate-resistant soybeans and Bt maize, which change the types of formulations involved in addressing post-emergent pesticide pressure.

Traditional agricultural crop pesticides can either be applied pre-emergent or post-emergent, a term referring to the germination status of the plant. Pre-emergent pesticide application, in conventional agriculture, attempts to reduce competitive pressure on newly germinated plants by removing undesirable organisms and maximizing the amount of water, soil nutrients, and sunlight available for the crop. An example of pre-emergent pesticide application is atrazine application for corn. Similarly, glyphosate mixtures are often applied pre-emergent on agricultural fields to remove early-germinating weeds and prepare for subsequent crops.

Pre-emergent application equipment often has large, wide tires designed to float on soft soil, minimizing both soil compaction and damage to planted (but not yet emerged) crops. A three-wheel application machine, such as the one pictured on the right, is designed so that tires do not follow the same path, minimizing the creation of ruts in the field and limiting sub-soil damage.

It is important also to note that even given appropriate chemical choices, high ambient temperatures or other environmental influences, can allow the non-targeted desirable organism to be damaged during application. As plants have already germinated, post-emergent pesticide application necessitates limited field contact in order to minimize losses due to crop and soil damage.

Typical industrial application equipment will utilize very tall and narrow tires and combine this with a sprayer body which can be raised and lowered depending on crop height. These sprayers usually carry the label 'high-clearance' as they can rise over growing crops, although usually not much more than 1 or 2 meters high.

In addition, these sprayers often have very wide booms in order to minimize the number of passes required over a field, again designed to limit crop damage and maximize efficiency. In industrial agriculture, spray booms 120 feet wide are not uncommon, especially in prairie agriculture with large, flat fields. Related to this, aerial pesticide application is a method of top dressing a pesticide to an emerged crop which eliminates physical contact with soil and crops.

Air Blast sprayers, also known as air-assisted or mist sprayers, are often used for tall crops, such as tree fruit, where boom sprayers and aerial application would be ineffective. These types of sprayers can only be used where overspray-spray drift-is less of a concern, either through the choice of chemical which does not have undesirable effects on other desirable organisms, or by adequate buffer distance.

These can be used for insects, weeds, and other pests to crops, humans, and animals. Air blast sprayers inject liquid into a fast-moving stream of air, breaking down large droplets into smaller particles by introducing a small amount of liquid into a fast-moving stream of air.

Foggers fulfill a similar role to mist sprayers in producing particles of very small size, but use a different method. Whereas mist sprayers create a high-speed stream of air which can travel significant distances, foggers use a piston or bellows to create a stagnant area of pesticide that is often used for enclosed areas, such as houses and animal shelters. In order to better understand the cause of the spray inefficiency, it is useful to reflect on the implications of the large range of droplet sizes produced by typical (hydraulic) spray nozzles. This has long been recognized to be one of the most important concepts in spray application (e.g. Himel, 1969), bringing about enormous variations in the properties of droplets.

Historically, dose-transfer to the biological target (i.e. the pest) has been shown to be inefficient. However, relating "ideal" deposits with biological effect is fraught with difficulty), but in spite of Hislop's misgivings about detail, there have been several demonstrations that massive amounts of pesticides are wasted by run-off from the crop and into the soil, in a process called endo-drift.

This is a less familiar form of pesticide drift, with exo-drift causing much greater public concern. Pesticides are conventionally applied using hydraulic atomisers, either on hand-held sprayers or tractor booms, where formulations are mixed into high volumes of water.

Different droplet sizes have dramatically different dispersal characteristics, and are subject to complex macro- and micro-climatic interactions. Greatly simplifying these interactions in terms of droplet size and wind speed, Craymer & Boyle concluded that there are essentially three sets of conditions under which droplets move from the nozzle to the target. These are where:

Sedimentation dominates : typically larger (>100 μm) droplets applied at low wind-speeds; droplets above this size are appropriate for minimising drift contamination by herbicides.

Turbulent eddies dominate : typically small droplets (<50 μm) that are usually considered most appropriate for targeting flying insects, unless an electrostatic charge is also present that provides the necessary force to attract droplets to foliage. (NB: the latter effects only operate at very short distances, typically under 10 mm.)

Intermediate conditions : where both sedimentation and drift effects are important. Most agricultural insecticide and fungicide spraying is optimised by using relatively small (say 50-150 μm) droplets in order to maximize "coverage" (droplets per unit area), but are also subject to drift.

Improved Targeting

In the 1970s and 1980s improved application technologies such as controlled droplet application (CDA) received extensive research interest, but commercial uptake has been disappointing. By controlling droplet size, ultra-low volume (ULV) or very low volume (VLV) application rates of pesticidal mixtures can achieve similar (or sometimes better) biological results by improved timing and dose-transfer to the biological target (i.e. pest). No atomizer has been developed able to produce uniform (monodisperse) droplets, but rotary (spinning disc and cage) atomizers usually produce a more uniform droplet size spectrum than conventional hydraulic nozzles. Other efficient application techniques include: banding, baiting, specific granule placement, seed treatments and weed wiping.

CDA is good examples of a rational pesticide use (RPU) technology (Bateman, 2003), but unfortunately has been unfashionable with public funding bodies since the early 1990s, with many believing that all pesticide development should be the responsibility of pesticide manufacturers. On the other hand, pesticide companies are unlikely widely to promote better targeting and thus reduced pesticide sales, unless they can benefit by adding value to products in some other way.

RPU contrasts dramatically with the promotion of pesticides, and many agrochemical concerns, have equally become aware that product stewardship provides better long-term profitability than high pressure salesmanship of a dwindling number of new "silver bullet" molecules. RPU may therefore provide an appropriate framework for collaboration between many of the stake-holders in crop protection.

Understanding the biology and life cycle of the pest is also an important factor in determining droplet size. The Agricultural Research Service, for example, has conducted tests to determine the ideal droplet size of a pesticide used to combat corn earworms. They found that in order to be effective, the pesticide needs to penetrate through the corn's silk, where the earworm's larvae hatch. The research concluded that larger pesticide droplets best penetrated the targeted corn silk. Knowing where the pest's destruction originates is crucial in targeting the amount of pesticide needed.

Application Methods for Household Insecticides

The first known aerial application of agricultural materials was by John Chaytor, who in 1906 spread seed over a swamped valley floor in Wairoa, New Zealand, using a hot air balloon with mobile tethers. Aerial sowing of seed has continued on a small scale.

Pest management in the home begins with restricting the availability to insects of three vital commodities: shelter, water and food. If insects become a problem despite such measures, it may become necessary to control them using chemical methods, targeting the active ingredient to the particular pest.

Insect repellent, referred to as "bug spray", comes in a plastic bottle or aerosol can. Applied to clothing, arms, legs, and other extremities, the use of these products will tend to ward off nearby insects. This is not an insecticide. Insecticide used for killing pests-most often insects, and arachnids-primarily comes in an aerosol can, and is sprayed directly on the insect or its nest as a means of killing it. Fly sprays will kill house flies, blowflies, ants, cockroaches and other insects and also spiders. Other preparations are granules or liquids that are formulated with bait that is eaten by insects.

For many household pests bait traps are available that contain the pesticide and either pheromone or food baits. Crack and crevice sprays are applied into and around openings in houses such as baseboards and plumbing. Pesticides to control termites are often injected into and around the foundations of homes.

Active ingredients of many household insecticides include permethrin and tetramethrin, which act on the nervous system of insects and arachnids.Bug sprays should be used in well ventilated areas only, as the chemicals contained in the aerosol and most insecticides can be harmful or deadly to humans and pets. All insecticide products including solids, baits and bait traps should be applied such that they are out of reach of wildlife, pets and children.

Aerial Application

Aerial application, commonly called crop dusting, involves spraying crops with fertilizers, pesticides, and fungicides from an agricultural aircraft. The specific spreading of fertilizer is also known as aerial topdressing.

Agricultural aircraft are often purpose-built, though many have been converted from existing airframes. Helicopters are sometimes used, and some aircraft serve double duty as water bombers in areas prone to wildfires.

Crop Dusting

The first known use of a heavier-than-air machine occurred on 3 August 1921 when as the result of advocacy by Dr B.R. Coad, a United States Army Air Service Curtiss JN4 Jenny piloted by John MacReady was used to spread lead arsenate to kill catalpa sphinx caterpillars near Troy, Ohio in the United States. The first commercial operations were attempted in 1924, by Continental Dusters which subsequently became part of Delta Air Lines. Use of insecticide and fungicide for crop dusting slowly spread in the Americas and to a lesser extent other nations in the 1930s and 1940s.

Top Dressing

Aerial topdressing, the spread of fertilizers such as superphosphate, was developed in New Zealand in the 1940s by members of the Ministry of Public Works and RNZAF, led by Alan Pritchard and Doug Campbell - unofficial experiments by individuals within the government led to funded research.

Initially fertilizer and seed were dropped together (1939), using a window mounted chute on a Miles Whitney Straight, but by the end of the 1940s different mixtures of fertilizer were being distributed from hoppers installed in war surplus Grumman Avengers and C-47 Dakotas, as well as some privately operated de Havilland Tiger Moths in New Zealand, and the practise was being adopted experimentally in Australia and the United Kingdom.

Crop dusting poisons enjoyed a boom after World War II until the environmental impact of widespread use became clear, particularly after the publishing of Rachel Carson's Silent Spring.

Water Bombing

Aerial firefighting, or water bombing, was tested experimentally by Art Seller's Skyways air services in Canada in 1952 (dropping a mix of water, fertilizer and seed), and established in California in the mid 1950s.

Night Aerial Application

Crop dusting at night is mostly liquid spray and is conducted in the Southwest U.S. deserts. The rising cost of pesticides and increasing immunity built up by continuous spraying reduced the effectiveness of spraying in daytime. In high temperature areas, the insects would travel down in plants in daytime and return to the top at night.

The aircraft - both fixed wing and helicopters - were equipped with lights, usually three sets: Work lights were high power and aimed or adjustable from the cockpit; wire lights were angled down for taxiing and wire or obstruction illumination; and turn lights were only turned on in the direction of the turn to allow safe operation on moonless nights where angle of entry or exit needed to be illuminated.

These aircraft were equipped with pumps, booms, and nozzles for spray application. Some aircraft were equipped with an elongated metal wing called a spreader, with inbuilt channels to direct the flow of dust such as sulfur, used on melons as a pesticide and soil amendment. Very little pesticide dust was used day or night in comparison to spray, because of the difficulty in drift control. Workers on the ground, called "flaggers", would use flashlights aimed at the aircraft to mark the swaths on the ground; later, GPS units replaced the flaggers due to new laws restricting use of human flaggers with some pesticides. According to the U.S. Bureau of Labor Statistics, in 2005 U.S. cropduster pilots earned an average annual wage of $63,210.

Environmental Issues

The aerial application industry is seemingly plagued with contradictory studies, some of which are negative and damaging and some are positive and show how the industry has evolved to become modern day environmental stewards. As with pesticide application in general, crop dusting is associated with a number of environmental concerns, including spray drift, soil contamination, water pollution, and occupational health concerns. Pesticide manufactures are continually evolving to address these concerns. In the infancy of the pesticide era little was known about the environmental impact of using pesticides.

Over time it became evident that better chemical compounds and more focused stewardship on behalf of the applicator were going to have to be researched and implemented. Recently, pesticide manufactures have developed new products that are designed to be safely metabolized when applied to prevent any unwanted or damaging environmental impacts; these new chemical compounds are referred to as "soft chemicals".

All crop protection products must meet tough safety standards. Only one in 20,000 chemicals actually survives the 8-10 year process of development, testing, and registration by the U.S. Environmental Protection Agency (EPA). Costs to test a pesticide's safety can range in price from $160 to $200 million. Nearly 900 scientists and program officials from the EPA make sure that products are properly registered to comply with federal law.

Once on the market, they are monitored by the EPA, the Food and Drug Administration (FDA) and state pesticide enforcement agencies. This stringent regulatory system ensures the safety of our food, the safety of the products to the environment, to water and to the farm workers that mix, load and apply the products. Another concern is that the aerial application of pesticides can lead to Pesticide resistance by insects, and is thus not a sustainable practice, however this is more to do with outdated farming practices and not the actual pesticides themselves. Modern day farming utilizes Integrated pest management (IPM) in order to reduce or eliminate pesticide resistance.

In the U.S. in 1970, lawsuits and court cases involving spraying of pesticides, such as aerial application in commercial agriculture were a growing area in law, combining areas such as negligence, products liability, strict liability, statutory regulation and commercial law.

Environmental and human rights issues associated with crop dusting is greatest in developing countries, where government oversight is weaker or absent, few safety practices are used, and chemicals are used that are banned in most developed countries.

This has led to many developments to mandate the proper dispersal of pesticides among all users. Aerial application specifically has improved equipment and techniques to provide accurate and effective delivery of crop protection products. In the 1960s the National Agricultural Aviation Association (NAAA) was established to foster industry development to the highest standards. Recently, NAAA developed the Professional Aerial Applicators Support System (PAASS) to educate pilots about safety, security and drift mitigation.

This is committed to reaching every aerial applicator in the U.S. with the latest information regarding these issues. In addition, the NAAA works with the federal government to invest in researching, developing and testing aerial application technologies to strengthen the safe application of crop protection products by air. Since the programs implementation statistical data suggests a substantial decreased to number of pesticide related incidences.

Crop dusting involving arsenic powders has been implicated in Bowen's disease. However, lead arsenic has not been used by aerial applicators or in any other form of agriculture for three decades because of the adverse affects to human heath that were not as well known when the powder was legal. There has been an ongoing debate about the effects of consuming food that has been treated with pesticides.

Often, the consumers receive conflicting data on this topic and find themselves in the crossfire of politically and emotionally charged facts pertaining to the pros and cons of pesticides.

The American Cancer Society states: "Many kinds of pesticides are widely used in agriculture in the production of our food supply. People who eat more fruits and vegetables, which may contain trace amounts of pesticides, generally have lower cancer risks than people who eat fewer fruits and vegetables. Pesticides play a valuable role in sustaining our food supply. When properly controlled, the minimal risks pesticides pose are greatly overshadowed by the health benefits of a diverse diet rich in foods from plant sources."

"Aerial application helps make it possible for us to have these fresh fruits and vegetables year round. Pesticide use has resulted in increased availability and a variety of low cost, fresh fruits and vegetables year round. This has had significant impact on human health because there is strong evidence that increased consumption of fresh fruits and vegetables is associated with a reduced risk of chronic disease, including many cancers.

Crop protection products also play a role in enhancing the safety of the food supply by reducing levels of natural toxins, such as mycotoxins and reducing the potential for contamination of fresh produce by food borne human pathogens".

Biodynamic Agriculture

Biodynamic agriculture is a method of organic farming that treats farms as unified and individual organisms, emphasizing balancing the holistic development and interrelationship of the soil, plants and animals as a self-nourishing system without external inputs insofar as this is possible given the loss of nutrients due to the export of food. As in other forms of organic agriculture, artificial fertilizers and toxic pesticides and herbicides are strictly avoided. There are independent certification agencies for biodynamic products, most of which are members of the international biodynamics standards group Demeter International.

The development of biodynamic agriculture began in 1924 with a series of eight lectures on agriculture given by philosopher Rudolf Steiner at Schloss Koberwitz in Silesia, Germany. The course was held in response to a request by farmers who noticed degraded soil conditions and a deterioration in the health and quality of crops and livestock resulting from the use of chemical fertilizers. At the course of eight lectures there were 111 attendees coming from six countries. An agricultural research group was subsequently formed to test the effects of biodynamic methods on the life and health of soil, plants and animals.

Regarded by some as the first modern ecological farming system and one of the most sustainable, biodynamic farming has much in common with other organic approaches, such as emphasizing the use of manures and composts and excluding of the use of artificial chemicals on soil and plants. Methods unique to the biodynamic approach include the use of fermented herbal and mineral preparations as compost additives and field sprays and the use of an astronomical sowing and planting calendar. Biodynamics originated out of the work of Rudolf Steiner, the founder of anthroposophy.

Today biodynamics is practiced in more than 50 countries worldwide. Demeter International is the primary certification agency for farms and gardens using the methods.

Geographic Developments

In the United States, the Biodynamic Farming & Gardening Association was founded in 1938 as a New York state corporation.

In Australia the first biodynamic preparations were made by Ernesto Genoni in Melbourne in 1927 and by Bob Williams in Sydney in 1939. Since the 1950s research work has continued at the Biodynamic Research Institute (BDRI) in Powelltown, near Melbourne Australia under the direction of Alex Podolinsky. In 1989 Biodynamic Agriculture Australia was established, as a not for profit association.

In France the International Federation of Organic Agriculture Movements (IFOAM) was formed in 1972 with five founding members, one of which was the Swedish Biodynamic Association.

The University of Kassel had a Department of Biodynamic Agriculture from 2006 to March 2011.

Germany, Italy and India are reported to be the leading countries in biodynamic agriculture based on biodynamic hectares.

Biodynamic Method of Farming

Biodynamic agriculturalists conceive of the farm as an organically self-contained entity with its own individuality, within which organisms are interdependent. "Emphasis is placed on the integration of crops and livestock, recycling of nutrients, maintenance of soil, and the health and well being of crops and animals; the farmer too is part of the whole." Cover crops, green manures and crop rotations are used extensively and the farms foster bio-diversity.

Biodynamic farms often have a cultural component and encourage local community. Some biodynamic farms use the Community Supported Agriculture model, which has connections with social threefolding.

Biodynamic Farmer

A biodynamic farmer perceives weeds and plant vulnerability to pests as a result of imbalances in the soil. Biodynamic agriculture sees the basis of pest and disease control arising from a strong healthy balanced farm organism. Where this is not yet achieved it uses techniques analogous to fertilization for pest control and weed control. Most of these techniques include using the ashes of a pest or weed that has been trapped or picked from the fields and burnt.

Pests such as insects or field mice (Apodemus) have more complex processes associated with them, depending on what pest is to be targeted. For example field mice are to be countered by deploying ashes prepared from field mice skin when Venus is in the Scorpius constellation.

Weeds are combated (besides the usual mechanical methods) by collecting seeds from the weeds and burning them above a wooden flame that was kindled by the weeds. The ashes from the seeds are then spread on the fields, then lightly sprayed with the clear urine of a sterile cow (the urine should be exposed to the full moon for six hours), this is intended to block the influence from the full moon on the particular weed and make it infertile.

Biodynamic Preparations

Steiner prescribed nine different preparations to aid fertilization which are the cornerstone of biodynamic agriculture, and described how these were to be prepared. Steiner believed that these preparations transferred supernatural terrestrial and cosmic "forces" into the soil. The prepared substances are numbered 500 through 508, where the first two are used for preparing fields whereas the latter seven are used for making compost.

A long term trial (DOK experiment) evaluating the biodynamic farming system in comparison with organic and conventional farming systems, found that preparations have influence on soil structure and micro-organisms enhancing soil fertility and increasing biodiversity. Regarding compost development beyond accelerating the initial phase of composting, some positive effects have been noted:

The field sprays contain substances that stimulate plant growth include cytokinins.

Field Preparations

Field preparations, for stimulating humus formation:

500: (horn-manure) a humus mixture prepared by filling the horn of a cow with cow manure and burying it in the ground (40-60 cm below the surface) in the autumn. It is left to decompose during the winter and recovered for use the following spring.

501: Crushed powdered quartz prepared by stuffing it into a horn of a cow and buried into the ground in spring and taken out in autumn. It can be mixed with 500 but usually prepared on its own (mixture of one tablespoon of quartz powder to 250 liters of water) The mixture is sprayed under very low pressure over the crop during the wet season, in an attempt to prevent fungal diseases. It should be sprayed on an overcast day or early in the morning to prevent burning of the leaves.

Both 500 and 501 are used on fields by stirring about one teaspoon of the contents of a horn in 40-60 liters of water for an hour and whirling it in different directions every second minute. Although, some biodynamic beliefs refer to buried quartz "fermenting", a 2004 review commented that it is unclear what this actually means, as rock does not ferment.

Compost Preparations

Compost preparations, used for preparing compost, employ herbs which are frequently used in medicinal remedies:

- 502: Yarrow blossoms (*Achillea millefolium*) are stuffed into urinary bladders from Red Deer (Cervus elaphus), placed in the sun during summer, buried in earth during winter and retrieved in the spring.
- 503: Chamomile blossoms (*Matricaria recutita*) are stuffed into small intestines from cattle buried in humus-rich earth in the autumn and retrieved in the spring.
- 504: Stinging nettle (*Urtica dioica*) plants in full bloom are stuffed together underground surrounded on all sides by peat for a year.
- 505: Oak bark (*Quercus robur*) is chopped in small pieces, placed inside the skull of a domesticated animal, surrounded by peat and buried in earth in a place where lots of rain water runs past.
- 506: Dandelion flowers (*Taraxacum officinale*) is stuffed into the peritoneum of cattle and buried in earth during winter and retrieved in the spring.
- 507: Valerian flowers (*Valeriana officinalis*) are extracted into water.
- 508: Horsetail (Equisetum)

One to three grams (a teaspoon) of each preparation is added to a dung heap by digging 50 cm deep holes with a distance of 2 meters from each other, except for the 507 preparation, which is stirred into 5 liters of water and sprayed over the entire compost surface. All preparations are thus used in homeopathic quantities. Each compost preparation is designed to guide a particular decomposition process in the composting mass.

One study found that the oak bark preparation improved disease resistance in zucchini.

The approach considers that there are astronomical influences on soil and plant development, specifying, for example, what phase of the moon is most appropriate for planting, cultivating or harvesting various kinds of crops. This aspect of biodynamics has been termed "astrological" in nature.

Seed Production

Biodynamic agriculture has focused on open pollination of seeds (permitting farmers to grow their own seed) and the development of locally adapted varieties. The seed stock is not controlled by large, multinational seed companies.

Trademark Protection of Term Biodynamic

The term Biodynamic is a trademark held by the Demeter association of biodynamic farmers for the purpose of maintaining production standards used both in farming and processing foodstuffs. This is not a trademark held privately in New Zealand and the trademark is intended to protect both the consumer and the producers of biodynamic produce. Demeter International is an organization of member countries; each country has its own Demeter organization which is required to meet international production standards (but can also exceed them).

The original Demeter organization was founded in 1928; the U.S. Demeter Association was formed in the 1980s and certified its first farm in 1982. In France, Biodivin certifies biodynamic wine. In Egypt, SEKEM has created the Egyptian Biodynamic Association (EBDA), an association that provides training for farmers to become certified.

Studies Farming

Studies have compared biodynamic farming methods to both other organic methods and to conventional methods. Most studies have found that biodynamic farms have soil quality significantly better than conventionally farmed soils but comparable to the soil quality achieved by other organic methods; the decisive factor is likely to be the use of compost. Studies of yields differ in their conclusions.

A 1993 study compared soil quality and financial performance of Biodynamic and conventional farms in New Zealand. The study reported that, "The Biodynamic farms proved in most enterprises to have soils of higher biological and physical quality: significantly greater in organic matter, content and microbial activity, more earthworms, better soil structure, lower bulk density, easier penetrability, and thicker topsoil." The biodynamic farms were just as financially viable on a per hectare basis. The study compared biodynamic farms with adjacent conventional farms, but didn't attempt to compare farms of similar size, or with similar crops.

A further study investigated whether biodynamic preparations had any effect on the yield and growth of lentil and wheat crops, weed populations and soil fertility in the short term. The study found that "[i]n general, soils and crops treated with biodynamic preparations showed few differences from those not treated". Plots tended with biodynamically treated compost produced results for yield, crop quality and soil fertility that were similar to those tended with non-biodynamic composts and NPK fertilizers. Some alteration was observed in the nitrogenous chemistry of the soil and grain where biodynamic field sprays were applied, however the study did not ascribe or discern any biological significance to the difference. Among the variables considered by the study, some measured outcomes correlated with biodynamic field spray usage, including a higher per-unit biomass yield ratio for lentils and a lowering of carbon and crude protein contents in wheat grains. The study's conclusion remarked that "any additional short-term benefits from biodynamic preparations remain questionable."

A long-term study conducted at a commercial vineyard in California compared vineyard blocks treated with biodynamic preparations alongside those tended with general organic farming methods, to examine effects upon soil and crop quality. "No differences were found in soil quality" during the first six years of the study, and analyses of other indicators including the yield per vine, clusters per vine, cluster and berry weight also showed there were no differences. The study did find a statistically significant (p-value < 0.05) difference in the yield-to-pruning weight ratio, indicating an "ideal vine balance for producing high-quality wine grapes" for the biodynamically treated crop, but noted the control vines had been "slightly overcropped". In one particular year of the study the biodynamically treated wine grapes had significantly higher Brix and notably higher total phenols and anthocyanins. In conclusion, the study found that biodynamic preparations "may affect" the vine canopy and chemistry, but showed no effects on the soil and tissue nutrient parameters measured in the study.

A 21-year study by the FiBL Institute in Switzerland compared the agronomic and ecological performance of biodynamic, organic and two conventional systems. The study found that nutrient input in the biodynamic and organic systems was 34 to 51% lower than in the conventional systems but crop yield was only 20% lower on average, indicating more efficient production. The total energy (for fuel, production of mineral fertilizer and pesticides, etc.) to produce a dry-matter unit of crop was 20 to 56% lower for the biodynamic and organic systems, and pesticide input was reduced by 97% (by 100% for the biodynamic system). In regard to soil aggregate stability, soil pH, humus formation, soil calcium, microbial biomass, and faunal biomass (earthworms and arthropods), the biodynamic system was superior even to the organic system, which in turn had superior results over the conventional systems. With the significant increase in microbial diversity in the biodynamic and organic systems, there was a significant associated decrease in metabolic quotient, indicating a greater ability to use organic material for plant growth.

Related

Ehrenfried Pfeiffer, a biochemist prominent in the early development of biodynamic preparations, developed a process for the bacterial conversion of municipal waste into compost usable in agriculture. The process was first used on a commercial scale in Oakland, California in the early 1950s.

In a newspaper editorial, Peter Treue argued that similar or equal results can be obtained using standard organic farming principles (which he also criticized as unproven in efficacy) and that the biodynamic preparations more resemble alchemy or magic akin to geomancy.

In a 1994 analysis, Holger Kirchmann concluded that Steiner's instructions were occult and dogmatic, and cannot contribute to the development of alternative or sustainable agriculture and that many of Steiner's statements are not provable because scientifically clear hypotheses cannot be made from his descriptions (for example, it is hard to prove that you have harnessed "cosmic forces" in the foods). Kirchmann asserted that when methods of biodynamic agriculture were tested scientifically, the results were unconvincing.

Further, in a 2004 overview of biodynamic agriculture, Linda Chalker-Scott pointed out that many of the research articles comparing biodynamics with conventional agriculture did not separate the use of biodynamic preparations from practices used in organic agriculture. The term "biodynamic" should not be used interchangeably with "organic" agriculture. Chalker-Scott concluded that "scientific testing of biodynamic preparations is limited and no evidence exists that addition of these preparations improves plant or soil quality in organically managed landscapes."

BIBLIOGRAPHY

- Arteca, R. 2015, Introduction to Horticultural Science, 2nd ed., Gengage Learning, Stamford, USA, p. 584. ISBN 978-1-111-31279-4
- "Why Horticulture?". Department of Horticultural Science.University of Minnesota.Retrieved 2 May 2019.
- Shyr, C.L. & Reily, H.E. 2017. Introductory Horticulture, 9th ed. Gengage Learning, Stamford, USA, p. 5. ISBN 978-12854-2472-9
- Harper, Douglas. "horticulture". Online Etymology Dictionary.
- Entry for yard Dictionary.com (presenting information supposedly from Random House Dictionary)
- Archived copy". Archived from the original on September 10, 2012. Retrieved September 21, 2012.
- Fullagar, Richard, Judith Field, Tim Denham, and Carol Lentfer (2006) Early and mid Holocene tool-use and processing of taro (Colocasia esculenta), yam (Dioscorea sp.) and other plants at Kuk Swamp in the highlands of Papua New Guinea Journal of Archaeological Science 33: 595–614
- von Hagen, V.W. (1957) The Ancient Sun Kingdoms Of The Americas. Ohio: The World Publishing Company
- Solomon, Dawit, Johannes Lehmann, Janice Thies, Thorsten Schafer, Biqing Liang, James Kinyangi, Eduardo Neves, James Petersen, Flavio Luizao, and Jan Skjemstad, Molecular signature and sources of biochemical recalcitrance of organic carbone in Amazonian Dark Earths, Geochemica et cosmochemica ACTA 71.9 2285–2286 (2007) ("Amazonian Dark Earths (ADE) are a unique type of soils apparently developed between 500 and 9000 years B.P. through intense anthropogenic activities such as biomass-burning and high-intensity nutrient depositions on pre-Columbian Amerindian settlements that transformed the original soils into Fimic

Anthrosols throughout the Brazilian Amazon Basin.") (internal citations omitted)

- Ancient society of York Florists,oldest horticultural society in world,longest running horticultural show in world established 1768".
- Australian Society of Horticultural Science – Australian Society of Horticultural Science".

INDEX

9 789360 849320

Printed at Repro India Ltd.